100% Utilization

Software Studies

Lev Manovich and Noah Wardrip-Fruin, editors

Expressive Processing: Digital Fictions, Computer Games, and Software Studies, Noah Wardrip-Fruin, 2009

Code/Space: Software and Everyday Life, Rob Kitchin and Martin Dodge, 2011

Programmed Visions: Software and Memory, Wendy Hui Kyong Chun, 2011

Speaking Code: Coding as Aesthetic and Political Expression, Geoff Cox and Alex McLean, 2012

10 PRINT CHR$(205.5+RND(1));: GOTO 10, Nick Montfort, Patsy Baudoin, John Bell, Ian Bogost, Jeremy Douglass, Mark Marino, Michael Mateas, Casey Reas, Mark Sample, and Noah Vawter, 2012

The Imaginary App, Paul D. Miller and Svitlana Matviyenko, 2014

The Stack: On Software and Sovereignty, Benjamin H. Bratton, 2015

Coding Literacy: How Computer Programming Is Changing Writing, Annette Vee, 2017

The Software Arts, Warren Sack, 2019

Critical Code Studies, Mark C. Marino, 2020

How Pac-Man Eats, Noah Wardrip-Fruin, 2020

Live Coding: A User's Manual, Alan F. Blackwell, Emma Cocker, Geoff Cox, Alex McLean, and Thor Magnusson, 2022

Thresholds of Digital Gameplay, Daniel L. Gardner, 2025

100% Utilization: Computation and Labor After Moore's Law, Andrew Lison, 2026

100% Utilization

Computation and Labor After Moore's Law

Andrew Lison

The MIT Press
Cambridge, Massachusetts
London, England

The MIT Press
Massachusetts Institute of Technology
77 Massachusetts Avenue
Cambridge, MA 02139
mitpress.mit.edu

The MIT Press would like to thank the anonymous peer reviewers who provided comments on drafts of this book. The generous work of academic experts is essential for establishing the authority and quality of our publications. We acknowledge with gratitude the contributions of these otherwise uncredited readers.

This book was set in Stone Serif and Stone Sans by Westchester Publishing Services. Printed and bound in the United States of America.

Library of Congress Cataloging-in-Publication Data

Names: Lison, Andrew author
Title: 100% utilization : computation and labor after Moore's law / Andrew Lison.
Other titles: One hundred percent utilization
Description: Cambridge, Massachusetts : The MIT Press, [2026] | Series: Software studies | Includes bibliographical references and index.
Identifiers: LCCN 2025030329 (print) | LCCN 2025030330 (ebook) | ISBN 9780262051361 paperback | ISBN 9780262051378 pdf | ISBN 9780262051385 epub
Subjects: LCSH: Moore's law | Technological innovations—Philosophy | Labor supply—Effect of automation on—Philosophy | Computer systems—Philosophy
Classification: LCC T173.8 .L57 2026 (print) | LCC T173.8 (ebook)
LC record available at https://lccn.loc.gov/2025030329
LC ebook record available at https://lccn.loc.gov/2025030330

EU Authorised Representative: Easy Access System Europe, Mustamäe tee 50, 10621 Tallinn, Estonia | Email: gpsr.requests@easproject.com

Unlike programs, computers must obey the laws of physics.

—Alan J. Perlis, foreword to Harold Abelson and Gerard Jay Sussman with Julie Sussman, *Structure and Interpretation of Computer Programs* (1985) 2nd edition (Cambridge, MA: MIT Press, 1996), *xii*.

Contents

Series Foreword ix
Acknowledgments xi

Introduction: The Ground Truth of Digital Mediation 1

1 **Approaching 100% Utilization** 23

2 **Economies and Scales: Producing and Consuming Computing Power** 45

3 **Squeezing Silicon: From Software Control to Hardware Efficiency** 69

4 **Convolute N(eural): Overused AI and Underemployed Minds** 101

5 **Winter in the Valley: The Accelerating Dynamics of Stagnation** 127

6 **Beyond Homeostasis** 163

Conclusion: Indeterminate Elements 189

Notes 203
Index 279

Series Foreword

Software is deeply woven into contemporary life—economically, culturally, creatively, politically—in manners both obvious and nearly invisible. Yet while much is written about how software is used, and the activities that it supports and shapes, thinking about software itself has remained largely technical for much of its history. Increasingly, however, artists, scientists, engineers, hackers, designers, and scholars in the humanities and social sciences are finding that for the questions they face, and the things they need to build, an expanded understanding of software is necessary. For such understanding they can call upon a strand of texts in the history of computing and new media, they can take part in the rich implicit culture of software, and they also can take part in the development of an emerging, fundamentally transdisciplinary, computational literacy. These provide the foundation for Software Studies.

Software Studies uses and develops cultural, theoretical, and practice-oriented approaches to make critical, historical, and experimental accounts of (and interventions via) the objects and processes of software. The field engages and contributes to the research of computer scientists, the work of software designers and engineers, and the creations of software artists. It tracks how software is substantially integrated into the processes of contemporary culture and society, reformulating processes, ideas, institutions, and cultural objects around their closeness to algorithmic and formal description and action. Software Studies proposes histories of computational cultures and works with the intellectual resources of computing to develop reflexive thinking about its entanglements and possibilities. It does this both in the scholarly modes of the humanities and social sciences and in the software creation/research modes of computer science, the arts, and design.

The Software Studies book series, published by the MIT Press, aims to publish the best new work in a critical and experimental field that is at once culturally and technically literate, reflecting the reality of today's software culture.

Acknowledgments

I began work on this project as a postdoctoral researcher in the digital humanities at the University of Kansas Hall Center for the Humanities. The editors at *Configurations*, Melissa Littlefield and Rajani Sudan, published my initial article on the topic, incorporated here in revised form. At the University at Buffalo, The Julian Park Fund in the College of Arts and Sciences covered costs associated with publishing the book. This allowed Shelby Brewster to carry out the indexing, for which I am particularly grateful. Drafting the artificial intelligence chapter was enabled by a UB Humanities Institute Faculty Research Fellowship, while the project as a whole was supported by the university's Office of International Education Manuscript Review Workshop in Global and International Research, in which Jacob Gaboury and Orit Halpern graciously agreed to participate. Others who commented on draft portions include Alexander Galloway, Matthew Ellis, Irving Massey, and Jonathan Cohen, as well as David C. Brock, who also ended up reviewing my proposal for the MIT Press. I would like to express my gratitude to him and the other, anonymous, reviewers, in addition to Noah Springer at the Press and the Software Studies series editors, for their efforts in bringing the book to fruition. Shintaro Miyazaki enabled me to preview some of the thinking here in an interview on minimal computing for his *Counter-N* project. He and Wolfgang Ernst also graciously allowed me to offer an overview of the book to the *Medientheorien* colloquium at the Humboldt-Universität zu Berlin. Portions were also presented at the Center for the Digital Humanities Seminar at the Göteborgs universitet, the University of Colorado, Boulder's Department of Critical Media Practices, and the Digital Scholarship Studio and Network at UB. I would like to thank Jonas Ingvarsson, Teri Rueb and Lori Emerson, and Cris Miller, respectively, for these invitations.

Many others helped in equally crucial ways: Jonathan Beller, Fred Turner, Matthew Fuller, and Anne Pasek all offered enlightening commentary on the book's themes. Wendy Hui Kyong Chun's invaluable guidance, over many years, inducted me into the debates to which this book responds and, more importantly, has supported me far in excess of them. Rick Prelinger's early supposition that I would connect the end of Moore's law with the declining rate of profit proved prescient, though I would not pursue this line of inquiry until much later. Louis Hyman and the 2021 Cornell University History of Capitalism Summer Camp inspired me to incorporate data visualization and financial reporting into the book's methods. At UB, the support and friendship of Alexander Reid, Carine Mardorossian, Tom Feeley, Paige Sarlin, Mark Shepard, Stephanie Rothenberg, and Matt Kenyon kept me on track. Across and beyond the walls of the academy, the camaraderie of Kyle Stine, Marc Böhlen, Timothy Scott Brown, and Michael Schulman has been a continual source of advice, relief, and merriment. Lastly, my family have been a limitless source of encouragement, inspiration, and understanding. I extend my deepest gratitude to them.

Introduction: The Ground Truth of Digital Mediation

In the 2020 limited television series *Devs*, a secret team of top engineers and scientists completes a quantum computing system for simulating the universe. Their program, after which the show is named, is total. It allows them access to any point in the past and, to a lesser extent upon which the series' climax hinges, the future. When, in the penultimate episode, the project comes fully online for the first time, Stewart, one of its key developers, reflects on the magnitude of the accomplishment: "A few hours ago we were in reality and we were working on a sim, and now we've pretty much traded. . . . The box contains us. The box contains everything. And inside the box, there's another box. Ad infinitum, ad nauseum." This fantasy of quantum omnipotence, however, this hyperbolic vision of computation's capacity to absorb the rest of reality, stands in stark relief to the performance of really existing twenty-first-century silicon computers.

Since shortly after the turn of the century, the clock rate of processors—the astoundingly large number of times per second that the staggering quantity of electronic switches within them can flip or flop—has remained largely stagnant. The reason for this has to do, paradoxically, with the very motor of progress that has made pop-culture visions of computer-simulated totality compelling to begin with. Intel Corporation cofounder Gordon E. Moore's "law" stipulates that the density—and, by extension, computing performance—of integrated circuits will double at the rate of, at most, every twenty-four months.[1] Since its initial formulation in the mid-1960s, the pace of development predicted by Moore, who passed away in 2024, has turned computation into a central component of contemporary life. By the mid-2000s, however, his dictate had resulted in these switches—transistors—becoming so small that the heat generated by their operation interfered with their neighbors.

For the last twenty years, this thermodynamic dilemma has meant that the halving of transistor size largely hasn't translated into the exponential clock speed growth of prior decades. Transistors have nevertheless continued to shrink, often in fits and starts, and electrical engineers have striven to find other ways to increase performance outside of raw cycles. However, we seem to be approaching a point where it is no longer just a question of overheating but of the physical impossibility of indefinitely continuing the process of miniaturization Moore's law describes.[2] The tech industry, in both its hardware and software engineering, is now struggling against the material boundaries of discrete silicon electronics. This book is thus situated between science-fictional dreams of unlimited computing power and the realities of digital mediation as an activity that always takes place in the physical world.

Perhaps the looming end of Moore's law is why our most outlandish speculations about the all-encompassing capabilities of computation have changed medium from integrated circuits to quanta. However, such a shift does not necessarily offer a solution to much of this dilemma. As one computer scientist puts it, "What does it mean to quantum-balance a chequebook?"[3] Instead, even as Moore's law unfolds into whatever lies beyond it, we are moving ever closer to a more fundamental limit: the extent to which real-world phenomena can be translated into equivalent artificial procedures.

While computational mediation's vertiginous delights may very well go on ad nauseum, then, they cannot do so ad infinitum. This fact poses profound problems not only for fictional fantasies of computational totality but also for a real-world political economy predicated on the replacement of labor with machinery. As a procedure enabling the mechanization of a wide range of phenomena, computation represents the most recent and extensive development of the instruments of production. If automation enables what remains of labor to be more productive, then stagnation in computing power poses significant problems for capitalism, reliant as it is on both economic growth and a lean workforce.

My own label for this development, a theoretical concept that responds to these conditions while at times going beyond the specific, historically determinate boundary of discrete silicon circuits, will be familiar to any programmer or network administrator from watching the output of system process monitors. It is the condition under which the capacities of a given

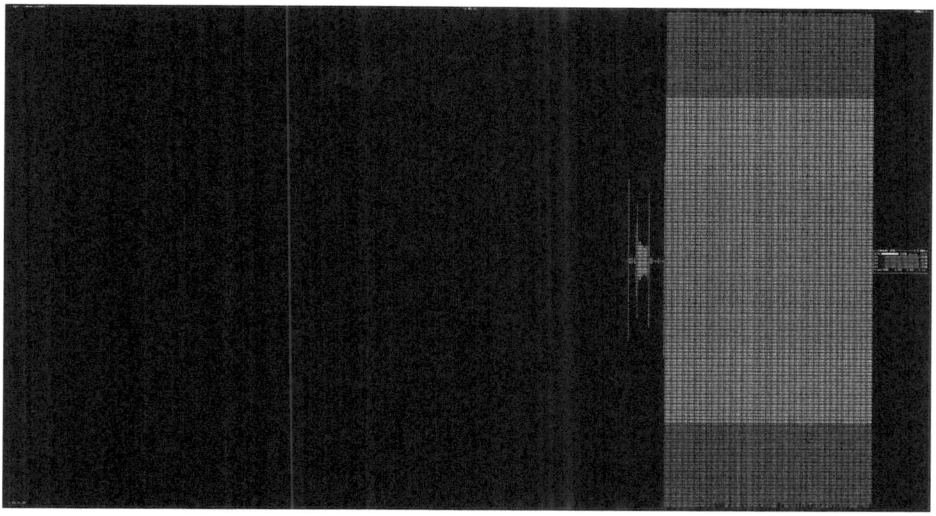

Figure 0.1
The system process monitor btop showing 100% CPU utilization on a Raspberry Pi 5.

processor—or collection of processors—are fully engaged, leaving no room for a system to do anything more: 100% utilization (see figure 0.1).[4] Its usage here, however, goes beyond this narrow technical definition and into the realm of the sociopolitical. Here it is not a matter of measuring a single system or group of systems. Rather, 100% utilization is my term for what happens when our global information society finds it can no longer rely on the future of exponentially increasing computing power it has come to expect. That day approaches even as we escalate our dependence upon computation, automating more and more labor away and turning a greater and greater number of physical objects into "immaterial" data streams in service, impossibly, of achieving a similarly exponential rate of financial return.

Responding to this intersection of the microelectronic and the macroeconomic thus requires the interpolation of disciplines not commonly thought together. Electrical engineering and computer science have, naturally, paid the most attention to the technical issues I am concerned with here. Marxist analyses, on the other hand, arguably offer both the deepest insights into the political-economic tendencies behind these developments and the most compelling frameworks for considering a post–Moore's law future. Far from representing a "break" with the capitalism of Marx's time, the development of information technology is deeply rooted in concepts like the structural tendency to replace labor with machinery, dependence upon technological innovation to do so, and the consequent repercussions he outlined. It is out of them that this book's deeper motivation arises.

Some approaches, beginning with Marx's own "Fragment on Machines" and extending through the autonomist and accelerationist approaches prominent in the first decade and a half of the twenty-first century, have emphasized the heightened contradictions of capitalism as it approaches an extreme of full or as-near-as-possible automation.[5] Here, the increasing sophistication of a "general intellect" externalized in machinery demands a reconceptualization of value on terms other than labor-time and, thus, a collapse of capitalist political economy. By contrast, other aspects of Marx's work, ones currently enjoying renewed interest, depict this mode of production as one of paradoxical, near-limitless expansion. There, cycles of industrial automation and overproduction result in falling profits and rising unemployment in relative terms but the inverse as measured absolutely.[6] The hope, in this critique, is that periodic upheavals in service of decreasing the ratio of employed to unemployed will eventually lead to a social revolution. The concept of 100% utilization synthesizes these two tendencies by acknowledging both the power of computation and our uncritical reliance upon it. In doing so, it introduces another, potential, equally transformative limit: the inability to reliably digitize past a certain point. My argument is that, since the conditions of seemingly endless growth to which we have become accustomed cannot continue forever, new ways of being in the world are not simply needed but imminent.[7]

100% Utilization in Context

Computer-scientific considerations of limits both logical and physical have tended to shy away from these considerations, even when solidly grounded in Marxist principles.[8] Engagements from a science and technology studies perspective have, on the other hand, emphasized the equitable distribution of computational capacity even while acknowledging that "chip manufacturing companies agonize over strategies to maintain" Moore's law.[9] Elsewhere within the humanities, other physical barriers to computational abundance have already been explored, most notably in terms of the cosmic timeframes necessary to generate the materials enabling it and the long-term environmental impact of its waste.[10] Meanwhile, infrastructural approaches to electronic mediation that have generated much attention in recent years have tended to emphasize nearly every element of our digital milieu aside from that of computation itself. Conduits, cables, cooling, and cellular access points are all crucial to both the centralization of high-performance computing power and the distribution of its effects, but processors themselves form a kind of "absent center" in this discourse.[11] Such materialist approaches to digital media find a precursor in at least one book from the explosion of new media studies in the century's first decade: Matthew Kirschenbaum's *Mechanisms*.[12] True to its moment, it focuses less on computation as such than its storage media. Newer considerations of computational limits in the field have tended to emphasize its logical barriers or the overall concept of incomputability.[13] Subfields that take up with integrated circuits, meanwhile, often do so from a historical standpoint bordering on nostalgia.[14] The retrospective character of these analyses, marveling at the capability of older hardware to do seemingly so much with so little, plays into a sense the "glory days" of computing are behind us, one that a looming end to Moore's law makes concrete. This feeling finds its counterpart in various critiques of a stagnating twenty-first-century culture, digital and otherwise.[15] Such diagnoses, in turn, connect with broader analyses of a global capitalism approaching an inflection point following a series of financial crises and political misadventures.[16]

It is my contention that these developments must be addressed from inside their connections to computation. Computerization has often been associated both temporally and conceptually with a shift in capitalism, at least in so-called advanced countries, from commodity manufacture to finance. However, it is a mistake to derive from this connection the view that computation itself has no real relationship to production in the classical sense. There are several reasons for this. Firstly, the distinction between mental and physical labor upon which a separation between stereotypically "postindustrial" activities (including finance) and manufacture often rests is not particularly legible, at least in the Marxist tradition. Marx characterized labor as "essentially the expenditure of human brain, nerves, muscles and sense organs,"[17] a definition that applies to the activities of the steelworker and data entry clerk alike, whatever else may distinguish them. Labor under an informational regime of production is still labor, and well in advance of any attempt to theorize it in its specificity.[18]

Secondly, more nuanced analyses of the role of financialization insist that it is a *part* of capitalism, rather than operating in some sense "outside" of it. Crucial to this understanding is Giovanni Arrighi's transposition of the circuit of capital, M–C–M′, from a schematic of accumulation, exploitation, and return into a periodic cycle of capitalist economic development.[19] In this reading, the latter moment is dominated by "phases of discontinuous change during which growth along the established path has attained or is attaining its limits, and the capitalist world-economy 'shifts' through radical restructurings and reorganizations onto another path."[20] Fredric Jameson has seized upon this framework to consider "the finance-capital moment of globalized society, the abstractions brought with it by cybernetic technology (which it is a misnomer to call 'post-industrial' except as a way of distinguishing its dynamic from the older, 'productive' moment)."[21] He argues that, in this conjuncture, "mass cultural production and consumption themselves—at one with globalization and the new information technology—are as profoundly economic as the other productive areas of late capitalism, and as fully integrated into the latter's generalized commodity system."[22] Computers are thus an aspect of production in every way, even if they come to prominence in a historical moment associated with declining returns. The question that haunts Jameson conceptually, here and elsewhere, is whether some alternate trail enabling a renewed rate of accumulation even exists today, as it did for Arrighi in previous rounds, or whether the limits capitalism now confronts are more absolute.

I propose to add to those breaking points more typically considered in Marxist analyses of late-stage capitalism—the lack of new territories for expansion and consequent increases in privatization, unemployment, and social unrest—the physical limits of the machines that have been responsible for the absorption of much of the energy, literally and metaphorically, once dedicated to industrial production. Indeed, a binary opposition between the two only holds insofar as we fail to recognize that computation itself is a physical operation. Marxist materialism has often been opposed to that concerned with the functioning of matter.[23] Where the latter has focused on relations between things, the former has emphasized relations between people. This is evident, on the one hand, in the first volume of *Capital*, in which "the social life-process" is equated with "the process of material production," and, on the other, numerous attempts, including relatively recent ones, to emphasize the intercourse of objects from a non-Marxist perspective.[24]

But the question of whether rocks can think, with all this might normally imply in terms of liberal ecumenism,[25] takes on a very different tenor when the speed of the silicon logic gates supporting an ever-growing range of economic activity—and determining relationships between fixed and variable capital—is at stake. The mechanization of computing, as formulated by Alan Turing, is straightforwardly about automation: the replacement of a human computer, a person whose job is to conduct calculations, with an object whose role is to execute computations. Both the rapid "dematerialization" of informational commodities like songs, films, and magazines into digital bitstreams as well as the ongoing, if slower, replacement of workers with robots in industrial manufacture are dependent on the physical properties of the machines we have built to approximate his vision. I argue that these two aspects of postmodern production, as well as the increasingly self-directed execution of securities transactions and other financial operations, should be considered, in a broad sense, forms of automation. They are all subject to the limits that an end to Moore's law presents.

The material particularities of the forces of production, in other words, are critical to the social order they mediate. This becomes especially apparent as we approach the limits that the physical constitution of our tools imposes upon their operation. This does not mean, however, that the limits to computing power are the same as those presented by environmental catastrophe. Between the barely conceivable timescales of hardware's geological origins and ecological afterlives, its vanishingly brief period of use is further shortened by the upheaval Moore's law has thus far perpetuated with each new generation of electronics. As I will explore further in chapter 3, an end to growth in computing power will have uncertain knock-on effects. It could prompt the development of even more power-hungry computing clusters as performance stops scaling.[26] Alternately, it could upend industries reliant on centralizing such resources in data centers as constant capital costs skyrocket. Here, it could even reduce waste and diminish electronic production, as the need to replace working systems with faster models would be obviated.

In this regard I am particularly keen to avoid an overly direct application of the concept of 100% utilization to the notion of degrowth. Degrowth and related concepts have of late become a cause célèbre both in ecocritical studies generally and Marxism specifically.[27] However, without endorsing all its precepts, I am more inclined to view the situation as Ståle Holgersen's ecosocialist approach does. That is, "Instead of focusing on 'more or less' growth, we should look in a different direction."[28] Like Holgersen, I stake a position that eschews and critiques both accelerationist celebration and ecologically motivated abnegation. The end of Moore's law represents an especially unique case of physical limit in that reaching it threatens to leave us no worse, at most, than where we are today, capability-wise. Limits to growth in computing power are less a signal to adjust ourselves to overall diminishing returns than one to seek out new, more egalitarian, avenues for progress.

The implications these boundaries hold for our era of digital globalization will dominate the rest of the book. To approach these realities without recapitulating the ideological mistakes of social-democratic materialism, however, it is first necessary to pierce through the pervasive cultural imaginary that equates computation with permanent revolution. The question of recursion raised by *Devs* is an ideal place to do so. It certainly demonstrates the bewildering and seemingly bottomless nature of digital mediation invoked by many commentators. Yet it also reveals the less apparent, yet very real, limitations to any such conception presented by the physical qualities of computational media.

Down the Technological Rabbit Hole

Devs' system architect Stewart's reflections on the newly recursive nature of his reality—that it is now fully computer simulated and that this simulation therefore itself necessarily contains another, simulated, computer running its own instance of reality, and so on—connect with a deep vein of critical and cultural theory. In it, the significance of this kind of recursion is traceable, if not back to its nominally analog manifestation in Jorge Luis Borges's tale of life-sized maps, then certainly to the media-inflected provocations of French philosopher Jean Baudrillard they inspired and, which, in turn (readers would be correct to sense a pattern here), would inspire the film *The Matrix*.[29] Despite its rather vulgar—from a Baudrillardian perspective—insistence on some kind of outside to manufactured totality, the Wachowskis' film made clear the applicability of these ideas to computerized worlds. In so doing, it rendered explicit what Werner Herzog's television miniseries *World on a Wire* and its 1990s Hollywood remake *The 13th Floor* were largely content to allude to. Far from being an artifact of the dot-com era, however, this dizzying way of conceiving computational mediation both precedes *The Matrix* and persists even as the 2021 iteration of the franchise ironically mocks its own absorption into the postmodern quagmire from which it sprung.

Over forty-five years ago, for example, Jean-François Lyotard referenced Borges's story to support his vision of a societal shift from classical, explanatory science to a paralogical one modeled on thermodynamic and quantum uncertainty.[30] Lyotard argued that stochastic language games corresponding to these physical frameworks have replaced empirical, deterministic observations such that justice depends on an equitable distribution no longer of material resources but electronic information. Notably, his solution required only "free access to the memory and data banks,"[31] not computing power itself. Today, on the other hand, smartphone and web applications have even more layers built into their operations than the mainframes Lyotard had in mind, implementing wrappers to web browsers and nesting other programming languages within their high-level JavaScript engines. On the backend, "virtualization" has similarly extended the concept of time-sharing, traceable back to the early 1960s, by consolidating what were once standalone servers into vast cloud infrastructures running multiple operating system instances on the same physical hardware.[32] This increasingly spiraling abstraction of computational mediation away from the physical world and toward the symbolic exerts a profound influence on analyses of the present-day tech industry. For example, while studies of fundamentally web-based systems have produced valuable insights into the drives and biases of contemporary Silicon Valley conglomerates, they are typically able to only speculate about the inner workings of these products even at the software level, let alone that of hardware.[33] This is the case because the algorithms and physical infrastructure supporting online applications, unlike their interfaces, are obscured from user scrutiny and often protected as trade secrets.[34] More speculatively, recent theorizations of the digital mode of production extrapolate from these conditions to envision a global computational supercontainer that reshapes symbolic and political order alike in ways both new and familiar.[35]

Recursion, of course, is a fundamental programming concept (alongside its counterpart, iteration). The problematics it introduces have been central to the development of digital media theory, particularly in its relation to continental philosophy. Alexander Galloway has most succinctly summed this aspect up, with the added bonus of evoking postmodern cinematic practices, as "*mise en abîme*: an outside that leads to an inside, which leads to another inside, and on and on."[36] Inside the box, in other words, there's another box. Galloway is giving name to a quality of digital encoding described in Lev Manovich's *The Language of New Media*, a foundational text largely preoccupied with these nested structures at the level of media files.[37] That such documents tend to serve as repositories for other media that are themselves vessels for yet other media is perhaps best indicated by the subsequently developed Matroska Multimedia Container format. Yet this aspect of digital mediation transcends any divide between abstract and concrete, informational and material, data and computation. It is equally applicable, as the foundational work of others like Friedrich Kittler and Wendy Hui Kyong Chun have made clear—the former in advance of Manovich, even—to programming languages, which proceed "downward" from human-comprehensible dialects like Inform and Processing to nearly unreadable machine code and beyond into the very physical arrangements at stake here.[38]

At the same time, aside from Kittler's early intervention, these and many subsequent scholars have largely focused their critiques at the level of symbolic manipulation rather than material operation.[39] Their analyses have illuminated many aspects of digital encoding and programming, as well as their connections to larger social dynamics and power formations. A materialist analysis of computation, however, can reveal how such formative moments in the study of digital mediation were themselves determined by historical conditions of unchecked computational growth. This environment has remained the default frame for much of our thinking about the subject, and much the same can also be said about those responsible for promoting these technologies. Jonathan Sterne, for example, notes that, in building a mid-1990s "proof-of-concept" player-on-a-chip capable only of rendering sixty seconds of MP3-encoded sound, the codec's engineers "assumed that increasing computing power and storage capacity would eventually allow such a player to be bigger and more powerful."[40] It is this context that the threat of 100% utilization brings into relief. Any speculation about the personal devices of tomorrow, let alone an omnipotent or supranational system of computation, will have to grapple with what it would and would not be possible for such material arrangements to do.

Central to this question is the debate on the nature of code, the level at which the symbolic and material operations of computation intersect, between Chun and Galloway. Originating in Galloway's assertion, riffing off a far more allusive statement from Kittler, that "code is the first language that actually does what it says," Chun responds that "this reduction . . . does not capture the difference between the various instantiations of code, let alone the empirical difference between the higher symbolic machine and the lower interactions of voltages."[41] That is, to run on a machine, source code is translated into machine code, and different languages, interpreters, and/or compilers will generate different machine code, even if two programs are written to carry out the "same" task. This machine code will, further, not exactly mirror the "source" that generated it since functionality easily expressed in higher-level languages must be translated into complex sequences of these more fundamental—and architecture-specific—instructions. Code cannot do what it says, then, at least not exactly, and that observation is central to the Derridean approach Chun levels against Galloway's initial literalism. This extends, for her, into hardware, or at least hardware schematics, because on the one hand "hardware does not need software to 'do something,'" since operations can be hard-wired (and enjoy a performance advantage from so being, as we will see in chapter 3), but on the other, the electronics diagrams that represent these material instantiations are themselves a form of abstract logic much like programming.[42]

It is in the gap between these last two contentions that a theory of 100% utilization can make its own intervention. Hardware does *something*, even if our *understanding* of what it does is a rationalist lure. Computation as a physical process, in other words, may not do exactly what its logical representations "say," but it inevitably *does what it does*. And what it does, specifically, has limits not only in terms of the symbolic disseminations that code's "various instantiations" proliferate but in terms of the material operations of computation themselves. Another way of putting this is that algorithms are evaluated not only in terms of *output* but also *speed*.[43] Both types of constraints, the semio-linguistic and the physical, can be made tangible by a simple—unscientific but illustrative—programming experiment combining the hierarchy of languages that conspire to obscure these limits with the very recursion at issue in foundational theorizations of digital vertigo.

Consider a simple countdown function that, when called, subtracts one from the number it has been provided and then recursively calls itself with this new number, over and over, until that number has diminished to one. A version of this program in Python might read:

```
import sys

r = 15401

sys.setrecursionlimit(r + 5)

def recur(x):
    if x > 1:
        x -= 1
        recur(x)
    else:
        print("Finished!")

recur(r)
```

In C, it could be written like this:

```
#include <stdio.h>

int r = 15401;

void recur(int x) {
    if (x > 1) {
        x--;
        recur(x);
    } else {
        printf("Finished!\n");
    }
}

int main() {
    recur(r);
    return 0;
}
```

That these functions may be roughly comparable at the level of code and apparently produce the same result but, in fact, operate differently is not a source of misrecognition here but precisely the point. The choice of 15401 as the number to be counted down from is not arbitrary. Rather, as the `sys.setrecursionlimit()` function called in the first snippet implies, this is just shy of the largest number of recursions that Python's internal stack, or register of active routines, can reliably handle on this system, a Raspberry Pi 4 with 8 gigabytes of memory, without crashing. True to the analyses discussed earlier, our recursive function is itself nested within the overall Python program, which in turn is encapsulated by the interpreter. A few additional recursions beyond the 15401 used in the logical act of counting down allow the program as a whole to complete. By comparison, a properly compiled C program, while also not limitless, can handle billions of recursions on the same hardware and thus count down from a similarly large value of `r`. These are logical, rather than material, restrictions. Using Python's lower limit in both instances provides an equivalent basis for comparison but is ultimately a functional boundary imposed by that interpreter and not the C compiler. The two languages—and more importantly, the machine instructions generated by each language's translator—differ.

This latter fact moves us toward a more physical limit. Though imperceptible to the naked eye, the variation in how long each program takes to run, as measured via the Linux `time` command on a cleanly booted system,[44] is substantial:

```
$ time python ./recur.py
Finished!

real    0m0.151s
user    0m0.078s
sys     0m0.057s

$ time ./recur
Finished!

real    0m0.011s
user    0m0.000s
sys     0m0.003s
```

Comparing the time each process takes to complete we can see that the C instantiation of the algorithm is substantially faster than its Python counterpart across all three metrics: total time elapsed from start to finish and central processor unit (CPU) time taken up by user- and system-mode activities. Both programs halt in a fraction of a second, though the specific amount of time elapsed will vary depending on the state of the system at the moment they are run. However, when scaled up from this relatively straightforward algorithm to the kinds of applications currently at the cutting edge of computing—virtual reality, codebreaking, or artificial intelligence, for instance—such quantitative differences become highly qualitative, affecting the utility of computation, even if they produce equivalent results. Not only do both languages have symbolic limits to their *mise en abîme*, then, but the amount of time they need to run makes certain things possible in the lower-level language that would take too long to realize in the higher-level one. In both instances, code does what it does, but in one of them it does so *much more efficiently*, and the extent of this efficiency is inversely proportional to the height of the programming-language hierarchy from which it descends.[45]

Computation, in other words, *takes time*. Yet the amount of time it takes depends not only upon the efficiency of the instructions to be executed, but first and foremost on the capabilities of the hardware doing so. It is reasonable to expect similarly *relative* differences in execution time between these two programs on any general-purpose system. However, their *absolute* speed is determined instead by the strength of that system's hardware. Programs impractical in a high-level language on low-end hardware may thus be workable—and even preferable, due to other factors like ease of understanding—on more powerful CPUs. The Pi 4 is a budget computer, and we should expect faster results as we scale up from there. However, as the looming end of Moore's law suggests, hardware cannot continue to get faster forever. Computational efficacy is therefore ultimately determined by the material efficiency of the medium upon which programs execute, and only secondarily by the economy of the algorithms it runs.

Visions of computation as an infinite cascade of boxes, then, run afoul of a problem that bedevils not only the programs examined earlier but thought experiments like Oxford philosopher Nick Bostrom's thesis that we live in a computer simulation being administered by a highly technologically advanced society, or, more provocatively and continuing with our opening theme, inside a computer simulation being administered by a highly technologically advanced society that is *itself* living inside a computer simulation being administered by a highly technologically advanced society, et cetera. Even if the amount of computing power necessary to do such a thing could be marshaled, he notes, the consequence of simulating systems inside systems would put an upper boundary on how advanced computers *within* these simulations could get: "One consideration that counts against the multi-level hypothesis is that the computational cost for the basement-level simulators"—that is, the real-world computers simulating however many layers of virtualized reality inside them—"would be very great. Simulating even a single posthuman civilization might be prohibitively expensive. If so, then we should expect our simulation to be terminated when we are about to become posthuman."[46] Ironically, just prior to this observation, Bostrom claims that "the metaphysical status of" whether "it is necessary for the hierarchy" of nested, simulated worlds "to bottom out at some stage . . . is somewhat obscure."[47] Its computational status, however, is decidedly not. Turing machines may be able to mimic the configurations of other Turing machines but only with a corresponding performance penalty. This concept, derived from computational biologist Michael Conrad, is central to Kittler's analysis of the relationship between software and hardware. It is, further, one inadequately addressed in most readings of the latter's famous essay, a situation this book aims to rectify.[48]

Overlooking or dismissing material determinants is a sure sign of ideological fantasy. Bostrom and *Devs*, not to mention rafts of tech evangelists entrenched in the recesses of Silicon Valley, rely upon notions of unlimited computer power to legitimate their greater goals and arguments. Such conceptions promise technological—and thus supposedly neutral rather than nakedly political—control of reality. This kind of maneuvering been previously noted, at least in passing. Cinema scholar Philip Rosen, for instance, in a critique of the novelty propounded by digital multimedia boosters in the 1990s, has argued that "it is necessary to insist that the physical characteristics of such infrastructures present limitations. Speaking very broadly, hardware limitations have to do with the speed of computation . . . but such considerations are neither historically nor theoretically determinant for the digital utopia, which perpetually forecasts more speed and more storage capacity."[49] Writing in this earlier moment, his focus is on the way that older models, particularly film theorists' conceptualizations of subjectivity, continue to hold despite proponents' claims for the newer medium's radical discontinuity. In the Lacanian–Althusserian framework leveraged by cinematic apparatus theory, "fetishism is a structure of disavowal" aligned explicitly with ideology, "and the digital utopia is shot through with disavowals," including "not only physicality, but . . . the materiality of the digital apparatus as well as the economics that guide its development."[50]

However, just as Rosen's focus on subjectivity, despite this acknowledgment, foregrounds the conceptual rather than corporeal requisites of the medium, his situation at the historical conjuncture during which the digital was arguably most aggressively cementing its dominance, chasing Moore's law for all it was worth, ends up vitiating some of his immediate critiques. His emphasis on the physical limitations of computation (and related infrastructural elements such as networks) derives from skepticism of "an imaginary of a world-wide network of unhampered informational flow."[51] Yet this attitude toward networks (as opposed to a world-wide *computer*) could seem anachronistic in an age when users in many if not most countries can access the same content from their phone while waiting at the bus stop as from their televisions at home and a liberal moral panic centered around the notion that "misinformation" is flowing *too easily* is in full swing. While many—though, of course, not all—forecasts from the 1990s have ended up as practical realities, the presently diminishing potential for computational improvement allows a fuller illumination of these material constraints. This, in turn, reveals the predictions of today's "digital utopians" to be even more baldly ideological than those of their predecessors twenty-five years earlier. Fantasies of unlimited or near-unlimited computing power can also misdirect our conception of determinism—specifically, the relationship between technological, economic, and individual influence on the physical world—in ways that will become clear by the end of this book. A theory of 100% utilization, then, confronts these visions with the concrete reality of information processing as a material process in an era of diminishing returns both computational and economic.

Computation and Labor After Moore's Law

In this sense, the looming end of Moore's law is not simply or even primarily a question of a crumbling ideological façade. Computing power is distinct from what we imagine its potential to be. It is no fantasy but a material fact. Academic computer scientists have often been remarkably clear-eyed about the inevitability of an end to growth in processing power, even as those who make their fortunes in the tech industry have regularly sought to disavow it.[52] This book is, instead, about the determining physical exigencies of computing and the reckoning in store when both they and the increasingly corresponding regime of financial growth to which they are currently shackled can no longer deliver, particularly in terms of their drive to replace people with machinery in the processes of economic production. It addresses ideological visions of processing power insofar as their critique reveals actual limits to what computation can achieve in the physical world. However, these very real boundaries make themselves felt independent of any ideas we may develop about them. Their implications for political economy are what I invoke when referring to 100% utilization.

That capital needs labor for its own valorization is the key insight of Marxist analysis. My consideration of discrete, electronic, silicon-based computing power in its twilight seeks to return to it via the indirect, but necessary, route that is the analysis of technology. Human labor and machinic work, though increasingly conflated, are not the same. A relentless drive toward automation has been the hallmark of capitalist political economy since its inception and frequently confronts us today in explicitly computational terms. Despite this, the creation of value is not reducible to the thermodynamic transfer of energy that causes engines, and now processors, to give off heat. Indeed, they are, in some small but important way, opposites.

To approach that question, however, we first must understand the methods by which the latter has come to replace the former in many domains and the consequences for us all when that process of substitution eventually peters out. As such, this book is divided into three parts of two chapters each. The first establishes the terms of analysis by considering the specific interrelationship between technology and economics that forms the basis of Moore's law, with each chapter leaning closer to one of these aspects. The second offers analyses of recent technological trends, focusing on the points where changes in hardware and software overlap with political economy. It concludes with a consideration of the drive toward—and limits to—synthesis between computation and the brain in the form of so-called artificial intelligence. The final section opens from there more directly upon the human and the social. It considers the influence of Silicon Valley culture and the ultimate question of the relationship between work as thermodynamic process and living labor as creative force.

Where this introduction has described the conceptual stakes of 100% utilization, chapter 1 provides the particulars. There, I lay out the contours of the problem represented by an end to exponential growth in computing power: the longer historical dialectic of labor and technology, the development of Moore's law from inception to stagnation, and the effects this has on hardware, software, and the economy. Stressing the heightening significance of material constraints on integrated circuits, its analysis juxtaposes Moore's law with capitalist dynamics of unemployment. This opening chapter thus puts computational growth into dialogue with both Marx's theory of the reserve army of labor and Kittler's heady and—as mentioned—often misunderstood consideration of computation. In doing so, it establishes the book's context of profound crisis, in which the human and technological components of global capitalism are, together, approaching a critical juncture.

Chapter 2 delves more deeply into hardware industry dynamics and the significance of computing power to our global information economy. It begins by examining the cost of manufacturing increasingly dense integrated circuits. While the expensive machinery necessary to produce ever-smaller components may give fabricators pause, I argue, the swelling payrolls of designers employed to deliver further performance at existing scales suggests that equipment cost does not pose nearly as substantial a threat to future processing improvements as a physical limit to density. At the same time, this danger is not isolated to the semiconductor industry but rather extends to the economy at large. Considering the role of computing power in business, finance, and platforms, I develop an expanded conception of automation to conclude that an end to Moore's law could prove simultaneously disastrous for capital and beneficial for humanity.

Chapter 3 takes up with the material conditions of digital technologies circa 2025. It focuses on the resurgence of custom and customizable hardware from a technical, rather than economic, perspective. Descending the computational stack from the graphical user interface and the software-driven general-purpose computer with its central processing unit, it considers two other types of chip, the field-programmable gate array (FPGA) and the application-specific integrated circuit (ASIC), which have received increased attention in the context of a waning Moore's law. FPGAs extend the logic of programmability into the material makeup of circuits themselves, allowing for processing power to be more directly allocated toward a concrete use case, like emulating a particular retrogaming console, without completely sacrificing the adaptability necessary to emulate a different one later. ASICs go further, hard-wiring algorithms that could be written in software permanently into silicon for maximum efficiency in whichever narrow set of operations their design targets. The net effect of these developments is a paradoxical situation in which applied computation, still generally understood as the direction of a "universal" processor, is returning to its roots in specialized circuitry to eke out the last remnants of available computing power. This carries implications for the balance between human and technological valorization exemplified in processing-intensive activities like cryptocurrency mining. Today, those seeking new Bitcoin have largely become dependent on ASICs' ability to provide the greatest computational advantage. Already running at 100% utilization in the sense of continually engaging their full capacities, maximally efficient systems like cryptocurrency miners are particularly susceptible to 100% utilization in the form of an end to Moore's law. This foreshadows looming issues with data center capacity in general and the environmental impact of computing. It is a vulnerability, I argue, that also provides an opportunity to reconsider the ends toward which our collective potential, both computational and otherwise, should be put.

Chapter 4 considers what is arguably the most trumpeted application for new hardware paradigms: artificial intelligence. Specifically, it examines the convolutional neural networks upon which machine learning and what passes for AI in the mid-2020s are built. I argue that the twenty-first-century vogue for these technologies is enabled, if not demanded, by the fundamentally parallel nature of simulated neuronal activity. More of a conceptual refashioning—and in that sense, a software innovation even as it leaves classical programming paradigms largely behind—neural-net algorithms have proven unexpectedly compatible with a third type of custom hardware: graphics processing units (GPUs). The software architecture of these systems, in other words, mirrors and thus maximally exploits the custom and multiprocessor hardware that has been developed in attempts to cope with the impact of Moore's law on single-processor clock speed. This, too, has unintended effects. It abandons a model at least nominally organized around imperative code, and thus programmer direction, in favor of one that "infers" features from large training sets without being explicitly told what to look for. This approach gives computation the appearance of lifelike sentience and raises, as many critics have now pointed out, a host of ethical questions. Yet it also finds its dialectical counterpart in the evacuation of intelligence from labor, beginning with the transformation of education, particularly higher education, into market-oriented skill training.

Moving from hardware and software to social issues, chapter 5 offers a consideration of 100% utilization's broader effects. Specifically, it traces the cultural impact of semiconductor development both as it stemmed from Silicon Valley and affects the region in the twilight of Moore's law. In the twenty-first century, the Valley has largely rejected its focus on electronics and hardware in favor of the software and services these material components enable. Returning to the birth of the transistor, I reflect on how this scientific breakthrough inaugurated not only a new technical regime of production but a correspondingly new ideology consummately suited to the "knowledge economy" it was bringing into being. This worldview, I hold, originated in the obsession with eugenics to which Nobel-winning semiconductor pioneer William Shockley applied his skills and background as his own business ventures began to peter out. An increasing emphasis on intellectual capability as a requirement for labor in an informational society ultimately transcends the explicitly racist ends to which Shockley ultimately pursued it in the form of postindustrial unemployment. What Shockley, who hired Moore along with seven other key scientists that would all leave his employ to found Fairchild Semiconductor (which Moore and another of the so-called traitorous eight, Robert Noyce, would subsequently depart to start Intel), saw as a new justification for racial hierarchy has since become a general condition of postmodern life. Its emphasis on intellectual performance underlies breathless fears of superintelligent machinery rendering humanity superfluous from Valley elites as well as propositions on the left to either embrace or deflate such a vision of the future. While those hyping such speculative developments disavow the precarious state of computing power growth, I argue, those who downplay them err by forswearing the economic contributions currently made by computation that an incipient end to Moore's law puts at risk. Synthesizing these two opposing positions, it concludes by gesturing toward what may reside beyond our increasingly stagnant techno-economic dynamics: a more harmonious relation not only among us, but between us and our machines.

Chapter 6 rounds out this analysis by exploring the fundamental conceptual—indeed, philosophical—question underlying the preceding examinations of semiconductor-based digital mediation and its cultural effects: the economic relationship between technology and people. The examples of artificial intelligence and computational singularity are only the latest indications that advanced machinery continues to challenge our self-conception, whether in real-world or speculative contexts. The labor theory of value, on the other hand, offers a sharp distinction between the two. In the classical Marxist framework, labor can *create* value, while machinery can only *transfer* it. The nineteenth-century thermodynamic laws behind the latter concept expanded in the twentieth century via the rise of cybernetics as a metadiscipline seeking to govern and even negate the entropic decay resulting from the transfer of energy in both machinic and human operations alike. This blurring between the two inaugurated a postindustrial order increasingly predicated upon electronic calculation and equally inclined toward a diminishing workforce. It also gave rise to an ambivalent critique via the musical genre of techno, originating from the deindustrializing Rust Belt. While techno reflected it, the end of Moore's law threatens to negate this cybernetic negation, imperiling the very regime of automated rationality designed to stave off such entropic failure. As our technological capacity for thermodynamic work plateaus, the unique characteristics of labor can thus rise to the fore once again. As they do, computation will cease to be the ceaselessly expanding domain it has been over the American Century and become instead something much more manageable.

One question that a potential rearticulation of the relationship between machines and human beings under conditions of 100% utilization brings to the fore is whether the end of Moore's law represents an absolute limit to computing power in general. Frequently, however, revolutions in social order have given rise to revolutions in science, and vice versa. Charles Babbage's vision of a mechanical computer, much like the one Turing later described, remained a pipe dream in the industrial era, but the invention of photography accelerated processes of dematerialization during the same period. It remains to be seen, then, if some alternative to Moore's law will reinstate exponential computing growth and, if so, whether such innovation is accessible to us under the present mode of production. The looming limit to computing power may be a physical fact of nature to which society will have to accustom itself, or a temporary obstacle that will be overcome either within or beyond a capitalist framework. While the second possibility seems remote given the explicit relationship between Moore's law and cost-effectiveness described in chapter 2, what is certain in all three cases is that 100% utilization presents an issue for capitalism to the extent the latter relies on computing power across financial calculation, industrial automation, and electronic mediation. I will return to this problem, which is ultimately a question about determinism, alongside the latest developments in computing, in my conclusion.

These considerations by no means exhaust the relationship between computing power and political economy, one that has been hidden behind the seemingly endless, but in actuality quite limited, appearance of layers inside layers, boxes inside boxes, and machinery in place of people. Yet, the time-sensitive nature of analyzing Moore's law as a process that changes qualitatively every few years—under previous circumstances at least, and perhaps for a short while longer still—has given some urgency to these reflections. It is my hope that this updated vision of the relationship between technology and labor will provide a framework within which others will be able to perform their own analyses. In doing so, may we all come to a better understanding of the role that we, and not just our creations, have to play in the history of the future.

1 Approaching 100% Utilization

Marx's formulation of the reserve army of labor and its relation to capital is well known. Unemployment is a necessary feature of capitalism because, despite being predicated upon the exploitation of labor, it paradoxically favors the concentration of fixed capital, particularly the machinic elements of the production process, over the concentration of variable capital, or human labor. As Ernest Mandel puts it, "If the extension of output maintained the given relationship between inputs of living labour and inputs of dead labour (machinery and raw materials), it would rapidly reach both a physical limit (the total available manpower) and hence a profit limit. Under conditions of permanent full employment, wages would tend to increase and erode profits to the point where capital accumulation and economic growth would gradually disappear."[1] Full employment, that is, would remove any restraint on wages, essentially undoing capitalism as a political-economic system. Capital's solution to this dilemma is thus to keep a section of the workforce in a state of unemployment, "a disposable industrial reserve army, which belongs to capital just as absolutely as if the latter had bred it at its own cost."[2] By restraining workers' ability to command the full value of their efforts, the unemployed ensure that capitalism can carry on without the risks presented by full employment.

This dynamic has typically played out in liberal societies with portions of the working-age population—some revolving, some permanent—relegated to an unemployment mildly cushioned by the welfare state. In recent decades, however, this situation has been breaking down along two fronts. Firstly, state support for unemployment has been attenuated via the spread of neoliberal austerity measures, removing many of the buffers historically put in place to ameliorate this condition. Secondly, real and projected increases in this reserve army's numbers pose a growing threat to the integrity of the capitalist system. These developments, alongside the barriers to expansion now inherent to capitalism's newly global scope, lead Fredric Jameson, in his reconsideration of *Capital*, to argue that "the structural unemployment in Marx's conception of the 'reserve army of capitalism,' once a secondary feature of this system, moves to the very forefront of its analysis today."[3]

That this shift in prominence, at least in postindustrial nations, is due not only to the movement of manufacturing, customer service, and technical work to cheaper regions but also to technological advances is not an insight reserved for those versed in Marxist political economy. It is, rather, a concern voiced with increasing frequency across the political spectrum. Late-2016 reports of a successful test run of self-driving long-haul trucks by the now-defunct Uber subsidiary Otto brought debates about the future of work—and working populations—in an age of increased automation out of academic discussions and into the mainstream media.[4] The vision such stories evoke is one in which the viability of capitalism is threatened not by overemployment but underemployment: "A real wage decrease has occurred even when these companies can't find enough people to drive trucks. What happens if even 10–20% of jobs are reduced by fully automated driverless trucks?"[5]

This situation is in some sense an inversion of a more familiar scenario in which high employment, and thus high worker compensation, drives the innovation of labor-saving technologies. Yet this dialectical relationship between people and technology, labor and automation, or the relations of production and the forces of production,[6] appears to have broken down along with the welfare state. Automation, and technological development more broadly, seems to be proceeding at its own pace regardless of the pressure put upon capital by wages. Our perception of automation, whether we consider its progress rapid or slow, is doubtless informed by the electronic quality of contemporary change. As many analysts have concluded, a computerized economy has less need for labor than an industrialized one. Sociologist Randall Collins, for example, has argued that "although Information Technology . . . generates new activities, it does not generate paying jobs at the same rate that it eliminates them."[7] Collins argues that the most relevant aspect of Marx's analysis is this dynamic, by which technology replaces human labor: Innovations in the forces of production, having decimated industrial labor, now come for the professional occupations long considered immune to computerization. This "stripped-down Marxism," in his view, is set to culminate as developments like artificial intelligence create massive joblessness via the "technological displacement of all categories of work by electronic machinery," proletarianizing new swathes of the workforce.[8]

Whether celebrated by Silicon Valley, castigated by an organized labor still recovering its footing after decades of decimation, or fretted over by economists and intellectuals, the notion that technological progress poses an imminent threat to what remains of the socioeconomic order developed over the last 175 years is being raised with increasing frequency. This fear has only grown since Collins's prognostications in the 2010s, as AI demonstrates increasingly sophisticated capabilities for symbolic manipulation amid a growing unionization drive within the white-collar workforce it threatens.[9] Yet I want to pose here a significant counterpoint to the discursive hegemony of technological development. What if the dialectical relationship between machinery and employment in advanced postindustrial societies has not broken down along the same lines as that between the people and their governments? What if it has, instead, shifted such that, while the lack of a reserve army of labor, that is to say, 100% employment, has historically posed little threat to postindustrial capitalism, the lack of a reserve of technological innovation, or 100% utilization, increasingly presents itself as a significant obstacle?

That this deficit poses a problem not only in terms of the general tendency toward unemployment but even for a postpandemic cycle of comparatively high employment reveals the centrality of this concern. This weak coupling of contemporary automation trends to those in employment can account for both how this discourse arose during a moment of growing unemployment as well as why the subsequent labor market recovery has yet to produce much in the way of commoditized technological alternatives despite the progress, hype, and fears surrounding AI. We might, then, be tempted to ask: Can employers automate much faster than they already are? To be sure, pressure to perpetuate this dynamic may well intensify if wages continue to rise in the short term. However, we are simultaneously approaching a much more foundational limit to the technology that has underwritten most of these labor-saving innovations over the past fifty years: the integrated circuit.

As the basic component of modern computers, and thus the driving force behind an increasingly extensive range of developments across innumerable fields as they have undergone digitization, improvements in silicon-based integrated circuit technology allow both computation and memory to keep pace with the increasingly complicated programs we write. Faster circuits ensure that CPU time, currently capable of executing a few billion instructions per second, is rarely fully employed. More importantly, they guarantee that this will continue to be the case as programs progressively become more elaborate.[10] Yet this situation, in which, in a mirror image of Marx's unemployed industrial militia, a technological reserve is maintained against the forces of computational complexity seeking to maximize processor utilization,[11] is rapidly approaching a material ceiling. The fulcrum on which it rests, the projection of ever-increasing semiconductor density known as Moore's law, has begun to falter. This situation, while the subject of a major scientific and engineering report drafted by the National Research Council (NRC) of the National Academies just a couple of years before Collins's argument, has rarely been considered from philosophical or cultural perspectives. That report addresses many of the same problematics elaborated here but primarily from the standpoint of necessary transitions in computer science and electrical engineering curricula, inquiry, and implementation. It takes up questions of socioeconomic dependence on technological progress only as a secondary matter. Yet, the authors argue, "a slowdown in the growth of computing performance will have global economic and political repercussions."[12] What follows is thus an exploration of both what it describes as the "fundamental physical limits of technology" and the broader humanistic ramifications they hold.[13]

There Is Hardware

Fairchild Semiconductor and subsequent Intel Corporation cofounder Gordon E. Moore's prognostications about the shrinking physical footprint of semiconductors have, as many observers have noted, taken on a life of their own not only within electrical engineering circles but also throughout broader understandings of technological development. First fully articulated in his 1965 article "Cramming More Components onto Integrated Circuits," Moore noted that semiconductor designs had a maximal component-to-cost ratio such that "the complexity for minimum component costs" in an integrated circuit "has increased at a rate of roughly a factor of two per year."[14] In other words, the most cost-efficient integrated circuit to make would double the number of components it contained every year. Moreover, in absolute terms, the cost of producing that ideal circuit would simultaneously decrease.[15] These developments led, as Moore predicted, to upheavals in digital technologies from computing to telephony, as well as amplification improvements in analog equipment like radios and stereos, even as he subsequently amended the periodicity of his analysis from one to two years.[16]

Given their low-level nature, there is some ambiguity regarding the extent to which Moore saw developments in integrated circuitry replacing human labor rather than existing technological processes. Either way, however, efficiency was very much on the agenda as he outlined his vision. "Integrated electronics will make electronic techniques more generally available throughout all of society," he wrote, "performing many functions that presently are done inadequately by other techniques or not done at all. The principal advantages will be lower costs and greatly simplified design."[17] As suggested by Moore's initial draft, in which the leading "r" is typed but crossed out for the tamer "evolutionary," the integrated circuit was a revolutionary commodity.[18] It promised to effect social change through its own internal dynamism. Among the "wonders" increasingly powerful chips would enable, Moore even included "automatic controls for automobiles."[19] As Moore himself put it in a fortieth-anniversary retrospective, "A series of incremental changes, over a sufficient period, results in dramatically different products."[20] Moore's law was the principle of the transformation of quantity into quality in material form.

The Hegelian aspects of what would eventually be known as "Moore's law" both precede its official inauguration and remain relevant in its twilight. Although he rejects the notion that it was "as has been claimed, an anticipation of the argument that Moore published five years later," as David C. Brock describes it, a 1960 presentation by subsequent networking and user-interface pioneer Douglas Engelbart at which Moore might have been present nevertheless laid out a similarly dialectical view of electronic development:

> Engelbart's argument was a simple syllogism, with two premises and a conclusion. His first premise was that the function of an electronic device depended on the relationship between a number of physical phenomena and properties. His

> second premise was that miniaturization would significantly change some of the physical phenomena and properties. Therefore, new designs for devices would be required to realize the same electronic function at this highly miniaturized scale, and novel physical phenomena at this scale would open the door to the design of entirely new kinds of devices.[21]

Engelbart's evocations of how shrinking electronics "will baffle the intuition and the understanding which has been developed in working with normal-sized devices," giving rise to "fresh phenomenoligical [*sic*] relationships at each new size scale for an investigation in terms of new devices," further call specifically to mind the dialectical process in the *Phenomenology of Spirit* by which the abstract understanding, or *Verstand,* is continually superseded, until their ultimate reconciliation, by higher levels of concrete experience.[22] The NRC report notes a similar dynamic: "As incumbent technologies are stretched to their practical limits, innovations are leveraged to overcome these limits. At the same time, they set the stage for a fresh round of incremental advances that eventually overtake any remaining advantages of the older technology. That technology-innovation cycle has been a driving force in the history of computer-system performance improvements."[23] That such dialectical observations about the effects of electrical engineering improvements should come from Engelbart makes explicit the connection between the software and UI advances he is most well known for and their material basis.[24] At the same time, the ongoing recourse to this dynamic renders equally clear the conception of progress upon which electronic digital computing—long considered to be of a piece with an antiteleological, postmodern outlook—continues to rest. Moore's law is thus the secret teleology connecting an industrial capitalism based upon ideals of linear progress with a postindustrial one that has supposedly abandoned such "grand narratives" for an ostensibly infinite, undirected unfolding of symbolic manipulation.[25]

Indeed, Moore's own formulation underlying the exponential increase in semiconductor density over the past fifty years has had an enormous impact not only on the technology sector but culture more broadly. In the 1970s, Paul E. Ceruzzi writes,

> Other metrics of computing performance were also increasing, and at exponential, not linear, rates. C. Gordon Bell of the Digital Equipment Corporation recalled that many engineers working in the field used graph paper on which they plotted time along the bottom axis and the logarithm of performance, processor

> speeds, price, size, or some other variable on the vertical axis. By using a logarithmic and not a linear scale, advances in the technology would appear as a straight line, whose slope was an indication of the time it took to double performance or memory capacity or processor speed.[26]

By the time computing was becoming an undeniable aspect of postindustrialized societies thanks to the popular adoption of the internet, this description had made its way into Moore's own interpretation. Transcending his initial, narrow focus on semiconductor density, in a 1998 article calling it "the most authentic version of Moore's law," Probir K. Bondyopadhyay includes a handwritten note from Moore, dated August 7, 1996. It reads, "The definition of 'Moore's Law' has come to refer to almost anything related to the semiconductor industry that when plotted on semi-log paper approximates a straight line. I don't want to do anything to restrict this definition."[27] As is fitting for a metric that seeks to quantify the rate at which the material basis for the digital incorporation of so much of the rest of the physical world will not only proceed but intensify, its applicability to electrical engineering at the dawn of the internet age became as general as that of the logic of digitization to media a few years earlier.[28] Moore's law as a cultural phenomenon, in other words, operates between the infinitesimally small concrete limitations of electronic component density and the broad rhetoric of technological innovation that originates in Silicon Valley but now pervades our globally networked, twenty-first-century discourse. It is thus important to consider its effects across both of these registers, and one of the few explicit engagements with computing performance within media studies provides an ideal place to do so.

Friedrich Kittler's essay "There Is No Software" is infamous among new media scholars for its provocative, titular argument, namely, that the very logic of computation, as defined by Alan Turing in 1936, militates against intellectual property. For Kittler, the entire point of Turing's universal machine is its ability to perform any computation. Thus software, as a list of which computations to be performed under what circumstances, is theoretically infinitely portable, whatever real-world limitations, such as differing processor architecture and peripherals or binary compilation, may exist at any given moment.[29] Less noted, however, is the extent to which Kittler's argument is conditioned by Moore's law, which makes its first appearance in the essay's third sentence, when he observes that "a computer memory's transistor cells . . . in the last three decades of Silicon Valley exploits, have shrunk to spatial extensions of less than one micrometer."[30] He goes on to make the case for a qualitative leap with the 1971 development of the Intel 4004 CPU, after which, he suggests, subsequent schematics were no longer drawn by hand but rather digitally designed via their predecessors, before arguing that, "in constructing the first integrated microprocessor . . . Intel's Marcian E. Hoff had given an almost perfect demonstration of a Turing machine."[31] With this statement an equivalence between the integrated circuit and Turing's hypothetical mechanism is drawn, upon which Kittler's argument against software is built.

By this short piece's conclusion, however, to which much less attention has been paid than its incendiary title, all three of these concepts are superseded. Drawing, as we saw in the introduction, on the work of Michael Conrad, Kittler argues that the benefits of discrete computation can only be realized alongside a significant penalty:

> Millions of basic elements work under almost the same physical conditions, especially as regards the most critical, namely, temperature-dependent degradations, and yet electrically all of them are highly isolated from each other. Only this paradoxical relation between two physical parameters, thermal continuity and electrical discretization on chip, allows integrated circuits to be not only finite-state machines like so many other devices on earth, but to approximate that Universal Discrete Machine into which its inventor's name has long disappeared.[32]

Crucially, this electrical isolation reduces the number of associations between elements compared to that of analog systems: "The growth rate of possible interconnections between these elements, that is, of the computing power as such, has proven to have as its upper bound a square root function. . . . Thus, the very isolation between digital or discrete elements accounts for a drawback in connectivity that otherwise . . . would be bounded only by a maximum equaling the square number of all elements involved."[33] This is in contrast, for Conrad, to "a physical system [that] can be structurally nonprogrammable and at the same time effectively programmable . . . a structurally nonprogrammable system that is computation universal."[34] A structurally programmable system, here, is one that replicates the discrete, procedural nature of programming in its material organization.[35] The "hypothetical, but all too necessary, type of machine" that is structurally nonprogrammable yet effectively programmable, Kittler argues, "would constitute sheer hardware, a physical device working amidst physical devices and subject to the same bounded resources" such that "software in the usual sense of an ever-feasible abstraction would not exist any longer" and "programming it will have little to do . . . with approximated Turing machines."[36] At the conclusion of Kittler's argument, somewhat unexpectedly, lies the return of the analog, and, with Conrad, a rejoinder to those who would characterize the difference between biological systems and discrete, logical ones as a potentially surmountable, and therefore immaterial, one.

Indeed, in the final paragraph of Kittler's essay, the famously antihumanist media theorist, in setting out the material limits of discrete computation, comes strikingly close to an endorsement of humanism:

> In what I have tried to describe, drawing quite heavily on recent computer scientists, as badly needed machines that are probably not too far in the future, certain Dubrovnik observers' eyes might be tempted to recognize, under evolutionary disguises or not, the familiar face of man. Maybe. At the same time, however, our equally familiar silicon hardware obeys many of the requisites for such highly connected, nonprogrammable systems. Between its million transistor cells, some million to the power of two interactions always already take place. There is electron diffusion; there is quantum-mechanical tunneling all over the chip. Technically, however, these interactions are still treated in terms of system limitations, physical side effects, and so on. To minimize all the noise that it would be impossible to eliminate is the price we pay for structurally programmable machines. The inverse strategy of maximizing noise . . . may well be the only way to enter that body of real numbers originally known as chaos.[37]

"There Is No Software" thus concludes with a paradox it is difficult not to read in terms of a classically dialectical formula still lacking its third, resolving term. Discrete systems are better at computation than analog systems but can never approach them in terms of the physical interactions they can account for. Analog systems, on the other hand, rife with physical connections, present greater difficulties for explicit programmability and accurate computation.[38] Like Conrad, Kittler preserves latitude for making computational use of either biological entities or the analog functions of silicon-based technology. His point is nevertheless clear: A computability to come will have to do away with the structures that define it today.

There Will Be No M(o)ore Hardware

The dictates of Moore's law thus form not only the starting point for Kittler's analysis of "our postmodern writing scene"[39] but also its eventual supersession. At the time "There Is No Software" was published, Intel processors were fabricated with a process at one micron or 1,000 nanometers in width, while they are now manufactured at less than 1 percent of that scale. However, instead of halving this size every two years as Moore's law dictates, from late 2014 until late 2021, the company's products largely ran on a 14 nm process.[40] During this period, Intel progressively encountered problems in mass-producing smaller-scale chips, introducing low-power laptop CPUs using a 10 nm process in 2018, but delaying more powerful desktop processors until 2021, and revealing that its subsequent 7 nm process—a "true" halving of its previous 14 nm node—would also be delayed.[41]

Despite these numerical values, fabrication process scales are less a firm physical measurement than a kind of approximation.[42] Transistors are therefore somewhat larger in practice than these names suggest. This, in turn, means that there is slightly more room for miniaturization before the material limits to Moore's law are reached. Taking advantage of this practice (or, according to them, "to provide a more accurate view of process nodes across the industry and to better reflect the balance of power efficiency, performance, and area in future nodes"), in 2023, the company renamed its 10 nm process to "Intel 7" and its 7 nm process to "Intel 4" in an attempt to match the marketing of its competitors.[43] These include the Taiwan Semiconductor Manufacturing Company foundry, Arm-based CPU designers like Apple, Google, and Qualcomm, and chip manufacturer/architect Samsung. As Intel itself increasingly comes up short against its rivals in both engineering and the marketplace, this maneuver aims to keep the feeling, if not always the reality, of biennial exponential improvement alive. While progress continues at the company practically synonymous with Moore's law, it no longer does so according to the latter's timetable.

Like laptop CPUs, however, the design of smartphone and tablet chips from these other companies tends to favor energy efficiency over performance. If we are concerned with the measure of available processing power, stronger CPUs remain the relevant metric, one whose rate of shrinkage has slowed dramatically, particularly since the mid-2010s. Obversely, if less powerful chips end up responsible for most workaday computing uses, this reflects diminishing expectations for the significance of raw computation, confirming the stabilization of a medium equally synonymous with novelty.[44]

Since debuting the 14 nm node, Intel also abandoned its "tick-tock model" of introducing new CPUs in alternating "process" and "architecture" stages—the former providing increased density and the latter logical refinement at existing scales—by adding a third, "optimization" step to this formerly dyadic schema.[45] (This earlier approach was also more dialectical, given the process stage represents a miniaturization of each new architecture and the architecture stage represents an optimization of each new process scale, much as Engelbart imagined.) Despite the reassurance offered, immediately following this announcement in the company's 2015 annual report, that "advances in our silicon technology have enabled us to continue making Moore's Law a reality," these and subsequent delays indicate difficulties in perpetuating the rate of improvement to which the semiconductor industry—and every other industry that relies on it, from computer and software development to automated driving concerns—have become accustomed.[46] This development was met with a declaration by the Semiconductor Industry Association that it would relinquish the role it had played, roughly since the time that Kittler's essay was written, in synchronizing endeavors to meet the conditions predicted by Moore. This declaration in turn prompted M. Mitchell Waldrop to write in *Nature* that "Moore's law, the principle that has powered the information-technology revolution since the 1960s, is nearing its end."[47] One major reason for this looming denouement is the inability of silicon to maintain its operational effectiveness at just two density doublings on from a scale of 10 nm (in real, rather than marketing, terms): "At that scale, electron behaviour will be governed by quantum uncertainties that will make transistors hopelessly unreliable."[48] Kittler's essay thus follows the trajectory of Moore's law itself, beginning with exponential and qualitative reductions in scale and ending in the nanoscopic tyranny of quantum physics.

Moore's law is not entirely at an end yet, and most of its proponents either deny that it is approaching any core limit or insist that this boundary, once reached, will precipitate a transference of exponential growth to some other area of technology. As Waldrop puts it, "Everyone agrees that the twilight of Moore's law will not mean the end of progress."[49] Frequently offered solutions to this impasse include switching from silicon to another base material that gives off heat at a much lower rate; moving the fabrication process from two to three dimensions to vertically arrange components; and embracing unconventional computing models based upon quantum physics or, perhaps making good on Conrad's and Kittler's claims, neurological processing.[50] Yet, as Waldrop also notes, such methods are largely unproven, sometimes only slightly more efficient, and rely on considerable breakthroughs yet to come.[51]

More utopian formulations, like the one offered by the British IT website *The Register*, suggest that "Moore's Law will live on, long after we've reached its physical constraints, as we explore the limits of creativity and imagination."[52] In some cases, the rhetoric of continual innovation and the shortcomings of potential alternatives even overlap. The final SIA report, for instance, argues that

> It is important to realize that despite the many announcements of the end of Moore's Law it is expected that the transistor density of future IC will continue to increase in the foreseeable future at historical rates.
>
> Even though producers of logic integrated circuits have been the most prominent champions of Moore's Law it is also true that *producers of memory devices have been the leaders in production of Integrated Circuits with the tightest tolerances and the smallest dimensions. In few words, they have been the real drivers of Moore 'Law* [sic] *and will continue to do so in the future.*[53]

Yet, increasing density via vertical fabrication, a possibility explored through technologies like Intel and Micron Technology, Inc.'s 3D XPoint on the memory side and Intel's Foveros for CPUs, is much easier for "memory chips, which do not have a heat problem: they use circuits that consume power only when a memory cell is accessed, which is not that often."[54] This differential accounts for some of the lead that a company like Samsung, primarily known in the semiconductor industry for its RAM and solid-state storage, can claim over companies like Intel. Yet, the interaction between components within a computer system is governed by an interrelated series of potential logjams in which, at any given moment, processors, memory, storage, peripherals, or user input could be waiting on any other of these elements.[55] Despite their smaller components, which translate into higher capacity, memory generally lags behind processors in terms of speed, meaning that processors can spend a nontrivial number of their cycles idle, awaiting information to process.[56] Technologies like 3D XPoint may thus succeed in doing away with the longstanding and oft-overlooked distinction between short-term memory and long-term storage that Wendy Hui Kyong Chun has argued is central to digital media's paradoxical combination of transience and permanence.[57] However, computing power, and hence the designation 100% utilization, remains the premier indicator of Moore's law's progress and exhaustion.[58]

As in Marxist theory more generally, approaching this elemental boundary has wider, superstructural effects. Most notably, during the long interregnum in which Intel proved unable to shrink its fabrication processes and continuing through its mid-2020s struggles, the perception of innovation within the computer hardware industry specifically, and the high-technology sector more broadly, has declined. Moreover, this has happened even as the rhetoric of innovation—and the fear of unemployment-inducing automation—has skyrocketed. For example, Apple, long treated as an industry bellwether for innovative personal computing, delivered a relatively disappointing product line in 2016. Its offerings, like the iPhone 7 and its revamped MacBook Pro, made headlines more for what they removed—in the former case, the 1/8″ analog audio jack, in the latter, all data ports save for those conforming to the then-new and still relatively uncommon USB-C form factor—than the features they added. In fact, Apple's reputation for "innovation" was at that point largely proceeding without reference to the core performance of its hardware. The latter product made use of a family of Intel processors known as Skylake, most of which, including those used in its top-of-the-line models, had already been on the market for over a year. Even prior to that, MacBook Pros had relied not upon the previous-generation Broadwell processors but those of the generation prior, Haswell, since 2013. Moreover, they went for over a year without any updates at all until the introduction of the Skylake models in late 2016.

Company detractors were quick to point out that these decisions, including the one not to postpone the already-belated new model for a few more months until the newest processors could be incorporated, represented a turn away from the focus on performance long associated with professional-class hardware.[59] Defenders, meanwhile, placed the blame for many of the laptop's shortcomings squarely on the shoulders of Intel, rather than Apple, foreshadowing the company's shift to its own, Arm-based CPUs across its entire product line.[60] Even with the introduction of these chips, high-end variants of the low-power architecture used in mobile devices and budget computers like the Raspberry Pi, the company has alternated between describing them in terms of "industry-leading performance" and "industry-leading performance *per watt*."[61] Despite their impressive performance under such conditions, as the latter descriptor indicates, ecology rather than efficacy has driven much of the discussion around Apple Silicon.

Particularly when extrapolated across a hardware industry that has done little more than follow in Apple's footsteps for the past fifteen years, these developments can be seen to form the material backdrop to what Florian Cramer has described as "the post-digital condition."[62] For Cramer, this situation describes the acceptance of digital media as a static and unremarkable, rather than novel and destabilizing, presence in contemporary cultural life. This has been accompanied by a shift in the high-tech economy away from the "revolutionary" hardware- and software-vending titans of previous generations such as Intel, Apple (Macintosh design team member Andy Hertzfeld's memoir is literally titled *Revolution in the Valley*[63]), Atari, Sun, Oracle, and Netscape and toward app- and service-based platforms like Google, Facebook, and Uber, which push continual, often insignificant improvements to a user base reared on free or subsidized products.[64] The contrast between this newer business model and the rhetoric of innovation that the technology industry has naturalized, thanks to Moore's law, as its own has become increasingly apparent. One need look no further in this regard than Anna Wiener's review of Alexandra Wolfe's late-2010s account of Bay Area technoculture. In her review, titled "It's Getting Harder to Believe in Silicon Valley," Wiener argues that "as tech becomes more pervasive, coverage that unquestioningly echoes the mythologizing impulse is falling out of fashion."[65] The experience of using the Valley's flagship products has similarly soured, even among those otherwise inclined to look favorably upon technological innovation. As xenofeminists Laboria Cuboniks have argued, "If 'cyberspace' once offered the promise of escaping the strictures of essentialist identity categories, the climate of contemporary social media has swung forcefully in the other direction, and has become a theatre where these prostrations to identity are performed."[66] The increasingly mainstream ubiquity of digital technology, however, is accompanied by a much less frequently noted trend: its decreasing dynamism. In this way, 100% utilization is not merely a cloud on the technological horizon in the form of a theoretical limit to Moore's law; it is very much an issue affecting computation presently.

The Revenge of Linear Time

Computation, at least as it has been conceived since Turing, has been understood at its core to be algorithmic, that is, a sequence of instructions geared toward the resolution of an objective.[67] Programming languages that mask this fact, such as declarative languages, nevertheless rely upon it, a situation encapsulated in Kittler's description of a "postmodern Tower of Babel" that "reaches from simple operation codes whose linguistic extension is still a hardware configuration, passing through an assembler whose extension is this very opcode, up to high-level programming languages whose extension is that very assembler. . . . What remains a problem is only recognizing these layers which, like modern media technologies in general, have been explicitly contrived to evade perception."[68] As we have already seen, this situation has proven generative for poststructuralist new media theory. Casey Alt likewise takes up with this model in his examination of "how . . . the tightly packed, internally consistent, lockstep logic of ever-faster calculations [got] pried apart to create a medium—a malleable material stratum for accommodating an incredibly diverse array of communications, expressions, and affects."[69] For Alt, the answer to this conundrum is the development of object-oriented programming (OOP) techniques and, subsequently, languages over the course of the 1960s and early 1970s.

Object-oriented dialects replace earlier programing language models based on sequential ordering (e.g., BASIC) or subroutines called from within a main function (e.g., C) with the logical separation of program components into individual entities, or objects. Objects contain their own variables (called "attributes"), functions ("methods"), and modes of communicating and invoking them ("messages"). As Alt describes it, "An object-oriented approach reverse-engineers complexity by accurately defining the specific algorithmic behaviors and properties of each of the discrete elements in the system such that an accurate simulation of complex behavior emerges, bottom-up, from the sum of individual interactions."[70] From the perspective of 100% utilization, there are two particularly noteworthy aspects of Alt's argument.

First, most of the developments he describes as material are in fact not material at all but rather interventions into the way code is structured, which is to say that they are abstract, semiotic innovations rather than concrete, physical ones. This is a development that Alt himself acknowledges at the third layer of his own attempt to map the computational Tower of Babel from the ground up:

> Given that all of the code produced by both procedural and structural programming in third-generation languages is ultimately compiled into either machine or assembly languages for the specific processor on which it is intended to run, it is important to note that the advances introduced by both procedural and structural programming are more about creating conceptual aids for the human programmer than fundamentally changing how the computation actually executes.[71]

Such caveats are less prominent in his considerations of subsequent layers, but they are not replaced with any more explicitly materialist claims. Rather—and this is the second key facet of Alt's thesis—the increasing conceptual sophistication of these layers ultimately effects a transformation from the sequential time of the CPU into an abstract space populated with digital subjects and objects akin to that described by French philosophers Henri Bergson and Gilles Deleuze.

> Thus, in opening up medium and movement, messaging [between objects] also opens up the possibility for Henri Bergson's concept of duration, in which time is experienced directly as a continuous global change within the relations and states of an entire field of objects. . . . This duration of object orientation is categorically different from the abstract time of procedural programming. Whereas time exists as a series of linear atomic elements in the earlier models of programming and computation, duration is the irreducible continuous unfolding and emergent web of interactions as a whole.[72]

The step-by-step nature of classical computation is thereby transcended and time is converted into space, part of object-oriented programming's contribution as "just one expression of epistemic cultural changes that characterized the decade of the 1960s," but which, in its similarities with Deleuzian philosophy, obviously resonates with both Kittler's characterization of computation, as well as many scholars' understanding of the 1960s, as quintessentially postmodern.[73]

I have written elsewhere on the ways in which the unsparingly quantitative technologies of digital mediation paradoxically give rise to theorizations of their effects along Bergsonian/Deleuzian lines.[74] Here I want to stress the interrelation of the two points highlighted above, namely, that object orientation's shift away from sequential time, which purportedly "exists as a secondary function of the number of instructions without any irreducible identity of its own," and toward a self-contained *durée* such that "the program feels more like an embodied community and less like a linear script,"[75] is largely *immaterial* in nature. In other words, these developments, and even the innovations that Alt highlights such as late binding, which maps variables as programs execute instead of upon compilation, do not ultimately change the fundamental nature of object-oriented programs as series of sequential steps executed by a central processing unit.[76]

Likewise, Alt describes encapsulation as a "process of deconstruction" in which "the splitting of a single computational thread into several distinct objects operating in parallel" occurs such that "the conceptual space of the program expands from the single line to multiple parallel lines, thereby opening up space from a one-dimensional line to a two-dimensional plane."[77] Despite this, parallel computation, which describes the ability of one or more programs to run simultaneously on multiple processors, is distinct from what, given Alt's focus on 1970s languages like Smalltalk, he seems to be describing here. This "splitting" is instead usually characterized as concurrency: the ability to separate elements of a program into individual processes, including for running on a single CPU, via task scheduling. This blurriness may be traceable back to its inventor, Alan Kay, who created Smalltalk in the 1970s at the Xerox Palo Alto Research Center and makes reference to "parallel models" and "parallel processing."[78] Efforts to actually parallelize Smalltalk, however, are relatively recent. Moreover, as we will see, they developed in direct response to the constraints on single-processor speed resulting from the increasingly tenuous viability of Moore's law.

Like Engelbart's, Kay's vision was buttressed by an expectation of rapid and large-scale improvements in single-processor computing power. Kay in fact claims he "immediately adopted many of [Engelbart's] ideas" after encountering him at the University of Utah in 1967, concepts which in turn, he wrote in 1993, made him think of Moore's law:

> I was almost frightened by the implications: computing as we knew it couldn't survive—the actual meaning of the word changed . . . there would be millions of *personal* machines and users, mostly outside of direct institutional control. . . . An *extensional* system seemed to be called for in which the end-users would do most of the tailoring (and even some of the direct construction) of their tools.[79]

In Kay's retelling, the word "parallel" appears only once, describing not computation but the ramifications of getting hardware designed to run the language working—that is, in relation to the impact of Smalltalk's existence in the physical world.[80] Further, Kay makes clear, in terms as Hegelian as

Engelbart's, that his hopes for this new programming paradigm were profoundly tied up with improvements in hardware. "Object-oriented design," he writes, "is a successful attempt to qualitatively improve the efficiency of modeling the ever more complex dynamic systems and user relationships made possible by the silicon explosion."[81] Quantity again begets quality, and it is the growing power of the discrete processor as postulated by Moore that makes "ever more complex dynamic systems" computable.

Yet it is this decrease in, or at least the stability of, the ratio between linear time and the ever-increasing number of instructions that software developers include in their programs that the winding-down of Moore's law has already threatened for over a decade. This situation reached a breaking point in 2004 as the raw clock speed of CPUs plateaued (see figure 1.1),[82] necessitating new approaches in order to continue the similarly exponential increase in computing performance which Moore's law had colloquially

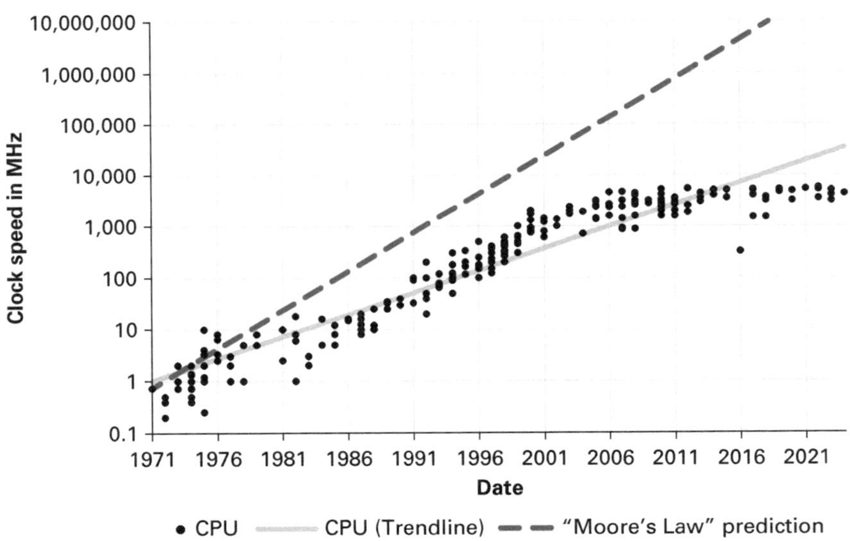

Figure 1.1
Clock speed of CPU models over time, beginning with the Intel 4004. Data adapted from the incomplete, but sufficient for illustrative purposes, list at https://en.wikipedia.org/wiki/Microprocessor_chronology, accessed March 9, 2025. CPUs listed without clock speeds excluded. Where a range of clock speeds was given, the highest was used. "Moore's law" prediction models doubling of clock speed every two years, not semiconductor density (hence scare quotes).

been understood to promise. Further reductions in scale were possible, but they would produce improvements in areas other than speed, such as power efficiency. Most notably, subsequent shrinking allowed hardware engineers to essentially place several processors in the same footprint as one of their predecessors, driving the software industry toward writing parallel code in order to improve performance.[83]

As software engineer Herb Sutter argued while this shift was taking place, refactoring code to take advantage of multiprocessor systems, while now a necessity in light of these "hard physical limits," can be a lengthy, intricate, and tenuous process: "The good news is that processors are going to continue to become more powerful. The bad news is that, at least in the short term, the growth will come mostly in directions that do not take most current applications along for their customary free ride."[84] Characterizing the necessary response to these conditions as the most significant paradigm shift since OOP, he notes that while compilers—and operating system software like Windows and macOS as well—can shoulder some of this burden, it is ultimately dependent on each program's developer to take full advantage of multiprocessing. Even so, Sutter concludes, "*Applications are likely to become increasingly CPU-bound*," rendering dubious Intel executive (and eventually, from 2021–2024, CEO) Patrick Gelsinger's contemporaneous postulation that "in the multiple core era, the exponential growth of computing performance will overtake the exponential growth of device complexity described by Moore's law."[85] Sutter's analysis was taken seriously in programming circles, even directly catalyzing work on the parallelization of Smalltalk.[86] We thus encounter a strange and paradoxical situation when comparing Alt's historical analysis to the contemporary moment. In the former, durational effects arise out of the largely conceptual reworking of programming paradigms around messaging between objects in the 1960s, one whose "validity," Alt himself concludes, "flows more from affect and belief than from mathematical correctness."[87] However, the recent, thoroughly material shift to a relational hardware structure based upon coordination between multiple cores, or even full CPUs, coincides instead with stagnation.

Thus, while many will maintain that Moore's law has a bright future, indications of its decline have already been felt for some time. It is therefore crucial for scholars of new media and cultural theorists alike to begin to reckon with both its historical hegemony and incipient exhaustion. This remains the case whether the exponential growth it predicts will end sooner or later, and whether or not it is ultimately superseded, as it almost certainly will be in the long run, by a new technology or metric. A theory of 100% utilization holds significance for a number of subdisciplines and scholarly preoccupations across the humanities. For infrastructure studies, which routinely considers technologies of storage and transmission such as network cabling and colocation facilities as distinct from technologies of computation, it collapses this opposition. As the NRC report notes (albeit in a slightly different context), "Switching and interconnect components . . . are part of every level of a computer system."[88] Similarly, the digital humanities has often relied on the notion that advances in computing power, rather than neoliberal shifts within the university, are responsible for its prominence.[89] The prevalence within the field of languages that by default only utilize one CPU, such as Python and R, suggests that this purported technological basis should be examined much more critically.[90] Environmental scholars, meanwhile, may note that the significance of material, physical restrictions on computing power reaffirms that the relationship between computation and nature is more complex than a simple narrative of opposition whereby the former represents a straightforward reification and exploitation of the latter.[91] Rather than strictly rationing it, an ecological response to the end of Moore's law could help realign computing power away from emulating real-world processes with ever-greater fidelity and toward those applications at which it excels.

As a concept, 100% utilization also challenges political interventions such as accelerationism, which, in arguing after Marx that "capitalism has begun to constrain the productive forces of technology," does not account for the physical, as opposed to social, constraints contributing to "only . . . marginally better consumer gadgetry," "relentless iterations of the same basic product," and "a world of minimal technical upgrades."[92] For fellow travelers like xenofeminism, which embraces technological change, particularly as it works to efface fixed notions of gender, a plateau in computing power may prompt further reflections on the differences between biological and computational systems and their interaction with each other, especially given its acknowledgment that "digital technologies are not separable from the material realities that underwrite them: they are connected so that each can be used to alter the other towards different ends."[93] It, finally, holds straightforward ramifications for scholars of new media, for, as Conrad argues, "Effective programmability is in general lost if all processors are used at once even if the machine is effectively programmable when operated in sequential mode."[94] We may thus eventually, as Chun suggests, "get beyond speed as the defining feature of digital media or global networked communications," but for material rather than intellectual reasons. Like capitalism under conditions of full employment, computational media under conditions of 100% utilization lose their dynamism.[95]

The model by which each digital layer encapsulates and is encapsulated by another—seemingly endlessly, but in fact, as Kittler has famously pointed out, down to the electrical current conditioned by the very semiconductors that are the focus of this book[96]—has rightfully gained much currency within new media theory. Yet it has also often been accompanied by a homology, sometimes explicit but often left implicit, between this model and epistemological frameworks drawn from poststructuralist theory, specifically those contending that there is no originary source of meaning or teleological endpoint, and certainly none with the kind of vulgar, material basis that Marxism has claimed as its own throughout much of its 170-year history. The looming end of Moore's law promises to rekindle the distinction between these two models, the former of which has always recognized in the electrical a "last instance,"[97] even if it has often been unable or unwilling to isolate it as such. Without the constantly increasing sophistication of our computing systems, the physical operations subtending them may gradually cease to be a source of mystery, rendering the computer closer to a hand tool than the arbiter of arcane instructions it is today. As Ceruzzi, echoing Hegel's invocation of the owl of Minerva, concludes, "Zeno's paradox states that we can never truly understand computers and computing, at least until innovation in the field slows or stops. Perhaps if Moore's law slows down, software writers will have time to catch up with cleaner and less buggy programming."[98] In a system that is no longer continually reinventing itself, as Sutter suggests, code will have to be engineered to meet a known material configuration as much as it can, just as hardware, as Waldrop and Mark Pesce note, is becoming more specialized in order to meet the needs of specific applications.[99] Despite a tendency to overlook or disavow the particulars of computation, its material constraints are becoming increasingly clear. While pundits of all stripes continue to fret over computerized automation, few, if any, have considered that the plateau represented by the end of Moore's law might be as dangerous to capital valorization as full employment.

2 Economies and Scales: Producing and Consuming Computing Power

If processing power functions as a reserve against rising labor costs, then what of the branch of the economy responsible for delivering it? How does the semiconductor industry operate? And what effects do its products have on the overall business environment? It is imperative to take these primarily economic considerations into account alongside the technological factors emphasized thus far. From the expense of fabricating next-generation silicon to the reliance of industry, finance, and platforms on Moore's law, they provide the context within which the threat of 100% utilization unfolds. It is to these two opposing perspectives then, the particularities of microchip production and the generalities of an economy that increasingly draws upon them, that I now turn.

It is necessary in the first instance to restrict ourselves to the semiconductor design and fabrication sectors, viewing them each as complementary halves of the industry writ large. In reality, just as designers depend upon fabricators to produce their chips, fabs must rely on producers like the Dutch ASML, which manufactures the advanced lithography equipment necessary to print wafers. After fabrication, chips are sent elsewhere for packaging and, usually, insertion into an end product like computers, add-on cards, or phones. Each of these antecedent and subsequent stages in turn has their own constituents, and so on, a consideration of which would obviously reveal additional inputs of both people and technology beyond those covered here. It would, further, highlight the substantial ecological costs associated with semiconductor production, which go far beyond the chemicals used in the fabrication process itself.[1] But it would also catch us in a recursive gesture that, if followed diligently, would encompass the entire economic output of the globe, after much consideration finding its terminus in the overall ratio between fixed and variable capital.[2] Taking the industry instead in relative isolation and following the contemporary, industry standard separation of its processes into planning and production allows us to better understand the relationship between machinery and labor within its boundaries. The microchip business is no less embedded in the global flows of capitalism than any other, and arguably more so. This, however, does nothing to change its drive to enable the computational simulation of these and other aspects of the real world as much as it can. The production of computing power is therefore of paramount importance. Yet the industry is simultaneously approaching the physical limits of Moore's law and struggling with the increasing cost of machinery capable of operating at ever-smaller scales. Perhaps surprisingly, these challenges—and the potential contradiction between them—can be illuminated via a fuller analysis of labor and its distribution across the layout–manufacture divide.

Rather than following the economics of chipmaking from here into every corner of global industry, it is more instructive to then "zoom out" and consider the impact of computing power more generally. It is possible to approach the same interrelationship of technology and political economy found in the specifics of circuit production instead from a vastly different vantage point. While economists have sometimes been skeptical of the bottom-line impact of computerization, it has clearly had a transformative effect on the ways in which business is conducted. Considering the range of its influence necessarily means further accounting for its tendency toward totalization. For capitalism, this propensity dovetails with its own need to continually extend itself into every aspect and moment of our lives. For computer scientists, its trajectory originates in the replacement of labor and moves into ever-more complex approximations of material processes requiring increasingly powerful capabilities. The capacity of integrated circuitry consumed globally for a growing variety of purposes thus reveals another way that labor is occulted, one that in turn necessitates further examination at the level of hardware and software.

Pursuing Performance: At What Cost?

If technology, as Fredric Jameson has repeatedly noted, is such a crucial aspect of Marxism that "it is very hard to ward off impressions of Marx as a technological determinist (and this in spite of his repeated insistency that value comes from human labor and not from machinery),"[3] then economics, as Moore himself is the first to admit, is also an integral part of technology: "Moore's law is really about economics. My prediction was about the future direction of the semiconductor industry, and I have found that the industry is best understood through some of its underlying economics."[4] For David C. Brock, "Moore's law is a description of human activity as well as a statement about the inherent possibilities of silicon semiconductor manufacturing technology. The law connects the work of people with the capabilities of silicon integrated circuit manufacturing through its focus on economics."[5] The ongoing debate over what will ultimately signal the end of Moore's law, a cost barrier to production at minute scales or the inability of the elements such fabrication relies upon to function at these levels,[6] is another indicator of how closely the two are connected.

While the concept of 100% utilization aligns more closely with the physical limits of computing, as classically formulated cost is a central factor in Moore's law. Since this is a more direct way into the economics of semiconductor manufacturing, it may seem surprising for a Marxist analysis to "take the long route" by delving instead into material barriers to infinite miniaturization. Certainly, one way to understand Moore's law is as naming the lowest point at which the balance between price and density can rest equally at a given moment in time. As Moore wrote in his initial formulation,

> Reduced cost is one of the big attractions of integrated electronics, and the cost advantage continues to increase as the technology evolves toward the production of larger and larger circuit functions on a single semiconductor substrate. For simple circuits, the cost per component is nearly inversely proportional to the number of components, the result of the equivalent piece of semiconductor in the equivalent package containing more components. But as components are added, decreased yields more than compensate for the increased complexity, tending to raise the cost per component. Thus there is a minimum cost at any given time in the evolution of the technology. . . . But the minimum is rising rapidly while the entire cost curve is falling.[7]

The difficulty of reliably producing increasingly complex circuits pulls against cost's overall tendency to decline thanks to the economy of scale.[8] Moore's law therefore can be seen as designating regular, progressive points of synthesis in which an ever-decreasing overall production outlay manifests even as the cost differential itself continues to flatten. Viewed this way, the more physical limits associated with shrinking semiconductor size hardly appear.

It is worth delving into the tension—if not contradiction—between these two outlooks. The antinomy is this: A cost barrier implies that component size is not the ultimately determining factor, while a material scaling floor designates a limit that no amount of capital can overcome. Approaching the question of cost backward, from the perspective of price, further illuminates some of the industry economics only hinted at in Moore's foundational article. To work from price to cost, in terms of both machinery and labor, let us consider what goes into pricing. Contrary to what Moore's law might imply, processor manufacturers do not offer a single chip at an optimal cost/complexity point. Rather, they differentiate their products.

Within a processor generation, prices typically vary according to features like clock speed, number of cores, and cache size. In Intel's modern lineup, for example, these are represented through designators like the Core 3, 5, 7, and 9 models (in order of increasing price and power). In addition, the company offers "K" versions of some chips, certified to operate at higher speeds than those for which they are officially rated, "F" versions missing an integrated graphics processor, and even "KF" versions combining these two aspects.[9] These offer additional points of pricing and market appeal. KF versions, for instance, are especially desirable to hardcore gamers, who chase performance yet use powerful external GPUs.

Perhaps surprisingly, however, these products are all generally versions of the same "blueprint," manufactured at varying levels of quality. High-end features are frequently present in lower-end chips, but inactive. Either they do not work due to errors in the manufacturing process or have not been tested. That is, the value of product differentiation here is not only in maximizing market segmentation but in minimizing losses due to defects in the production process. Instead of throwing away imperfectly fabricated products, such "binning" enables companies to functionally increase yield—and thus save on cost—by allowing less-performant chips to market.[10] Thus, while there is not a singular price on the consumer market for any processor generation, there is, in a sense, a uniform cost. From a production perspective, binning allows this cost to average out across yield. From that of consumption, the picture is less of one specific price than a range across which each wave of silicon differentiates itself.

This is one way that, in practice, the minimum cost to produce a circuit rises, since each production run incurs the same cost, no matter the quality yielded. But it is also a way to keep overall cost, and prices, lower than if chipmakers were to discard all but their most faultless commodities. As both Moore's law in its ideal formation and the practical realities of the CPU market suggest, then, cost does not set an *absolute* but rather a *relative* barrier to performance. This holds true as much for production as it does for sales. Just as consumers may elect to spend more or less on a CPU—that is, on a level of product quality—manufacturers may choose to invest to a greater or lesser extent in developing one.

Examples at the extreme end of the cost curve include the "wafer-scale" processors designed by artificial intelligence supercomputing firm Cerebras. Typically, multiple chips are printed on a single wafer and then cut out of it as individual dies. However, citing the same issues with perpetuating Moore's law discussed in previous chapters, Cerebras wrote on its blog that "since we won't be putting ever more transistors onto each square millimeter of silicon, we will instead put more square millimeters of silicon onto each chip."[11] This is one way that, as density gains peter out, performance may still continue to increase for a while, albeit at greater cost.[12] Given the same likelihood of defects in a fabrication process, the probability that any individual chip will be faulty is reduced by a factor of the number of dies the wafer is cut into.[13] It is therefore dramatically increased when using the entire wafer for a single design.

Cerebras addresses this via a technique similar to binning. It overproduces the number of identical processor cores needed across the wafer, enabling it to reroute the connections between them away from defective ones.[14] This lowers cost by reducing the risk that the whole wafer will have to be discarded due to a fault in one area. Still, the use of an entire wafer substantially expands the number of components on the chip. This in turn raises the cost, just as Moore described. The datasheet for the third iteration of their process, the WSE-3, introduced in 2024, claims four trillion transistors compared to eighty billion in an unnamed "largest GPU."[15] (Commodity CPUs can use even fewer.) Cerebras does not list its hardware prices and, in any case, sells complete systems, not individual chips, but each of its processors is estimated to cost between two and three million dollars.[16] The highest-end individual GPUs, by comparison, run into the tens of thousands, while CPUs typically weigh in around the mid-hundreds. Yet

Cerebras's gains in transistor count and performance come without increasing overall component density, rendering somewhat dubious their claim to have made "The Chip That Broke Moore's Law."[17] Rather, this formulation ignores two of its fundamental characteristics, cost and density, in favor of component count. Given the expense of doing so, it is perhaps not surprising that Cerebras argues "Moore's Law is likely to die an economic death before it hits the ultimate physical limits."[18]

The idea behind "an economic death" to Moore's law is that future generations of machinery would be too expensive to maintain the cost advantages historically associated with increased density. There are some indications that such calculations are already being made, with major fabricator TSMC resisting adoption of ASML's 2024 models. At the same time, TSMC Vice President of Business Development Kevin Zhang's statement on the matter, that "I like the technology, but I don't like the sticker price,"[19] sounds closer to a negotiating position than a nail in the coffin of economic viability, particularly as his company continues to push the density envelope forward. It is difficult to say if or when equipment costs will come down in the future as suppliers themselves continue to innovate, which they are obviously motivated to do. Economic barriers may therefore break the rhythm of Moore's law. Unlike the laws of physics, however, whether they spell a complete halt to increasing densification is less certain.

That density is a more absolute barrier than cost is true even within the economic boundaries set by the semiconductor industry. Chip designers cannot really work to shrink component size themselves. This task remains the domain of experimental research, largely aimed at developing commodifiable manufacturing procedures rather than commoditized products. Instead, as Cerebras's example shows, designers must operate within the process scales offered by manufacturers. This brings us to a topic that is painfully obscured in classical formulations of Moore's law, yet crucial from the Marxist perspective: the role of labor in the creation of integrated circuits. As we shall see, manufacturing labor is in fact a relatively small part of the overall cost profile of semiconductor production. Wafer-scale processors require extensive design investment—more than commodity chips, which themselves require an increasingly larger portion as transistors shrink.[20]

More broadly, the increasing necessity of specialized computing circuitry to prolong performance gains, the focus of the next two chapters from a technical perspective, can here be understood economically as driving demand for labor. Cost thus includes not only the expense of fabrication but that of developing a chip. As companies attempt to build out more functions at a given scale rather than wait for a smaller one to automagically solve performance problems, this requires a greater number of hardware designers. In Marxist terms this growth represents a threat that industry must seek to tamp down through further technological advancement. To do so would of course be significantly more difficult if and when the physical limits to semiconductor manufacturing are reached.

However, capital is very good at driving down manufacturing costs, especially when its livelihood is threatened. This is one reason an economic end to densification (though not Moore's law's already-shaken cadence) should be taken with a grain of salt. At the same time, as companies wait longer for scales to go down, industry-wide shifts in the distribution of labor and machinery are not simply indicators of short-term reconfiguration while the economics of a new generation of manufacturing equipment play out. They are signs of a trend that can only grow as we approach the material boundaries of semiconductor engineering.

While always two distinct moments in the overall production process, the differentiation of design from manufacture continues to direct transformative changes in the industry, particularly in the allocation of labor.[21] Semiconductor fabrication is notoriously capital-heavy. According to one US government report,

> As these fixed costs of production have risen, only a few firms can afford to build and maintain new wafer facilities, and many semiconductor firms have outsourced front-end fabrication to specialized foundries, such as Taiwan's TSMC. This trend has increased the popularity of firms that focus solely on semiconductor design work (fabless firms), allowing many semiconductor firms to shift risks of ever-higher fixed costs of production to specialized foundries.[22]

The narrowing of focus accompanying this segmentation has given companies like TSMC a manufacturing advantage over those such as Intel, who are neither fabless nor strictly manufacturers but "integrated device manufacturers" (IDMs) performing both functions.[23]

In addition to the distinctions between memory and processor density discussed in the previous chapter, this situation further accounts for Intel's relegation behind the vanguard of Moore's law. TSMC's process nodes are at least a generation ahead of the company Moore founded, and their resources are very carefully allocated. Perhaps surprisingly, the

WSC-3, whose manufacture is outsourced to TSMC (though the custom nature of their chip does require Cerebras to innovate in tooling), is not fabricated via the foundry's most advanced techniques, instead using their 2020 5 nm process.[24] Apple, another client, has sought to monopolize the company's 2022 3 nm technique, and has staked a large claim on its 2 nm production capacity as well.[25] Despite this, even Intel itself has been outsourcing some manufacturing to TSMC.[26] The company has also introduced accounting changes to distinguish its manufacturing and design operations.[27] It has thus acceded to the prevailing tendency to treat these two aspects of production as separate businesses, even while trying to keep them in-house. Moreover, with the late-2024 departure of CEO Patrick Gelsinger, this latter strategy could itself be in question.[28] Meanwhile, TSMC "has a significant global lead in ever smaller processes," arguably placing it in closest proximity to the physical limits of semiconductor engineering.[29] At the same time, as more fabless firms and even IDMs like Intel come to rely upon it, it is also the case that the limits TSMC confronts are increasingly those faced by the industry overall.

While such segmentation allows for increased specialization, and, therefore, performance, a financial comparison of foundries and fabless companies reveals stark differences in the balance between machinery and labor in each enterprise. In 2023, TSMC listed direct labor outlay at NT$22,193,265,000, or roughly $676 million, in comparison with a gross machinery and equipment valuation (not including office equipment or that under installation) of NT$5,131,811,792,000 (~$156.4 billion).[30] Even accounting for NT$99,950,197,000 (~$3 billion) of additional labor cost classified similarly but not broken out in the cost of revenue statement, as well as an additional NT$57,108,147,000 (~1.74 billion) in research and development payroll,[31] which bring total labor costs we can associate with design and manufacturing (but not other categories of labor like administration and sales) to NT$179,251,609,000 (~$5.46 billion), this yields a ratio of variable to fixed capital of just under 3.5%. While the categories are slightly different, and comparison is additionally exacerbated by divergences in international accounting procedures, the fabless graphics (and now artificial intelligence) company NVIDIA's ratio is so astronomically higher as to leave no doubt as to the overall relationship between these two industry segments. An exemplar in the space rather than an outlier like Cerebras, the company does not enumerate labor directly but does list R&D operating expenses of

$8.675 billion alongside a gross equipment, compute hardware, and software valuation of $5.2 billion over roughly the same period.[32] This rough calculation yields a ratio of just under 167%. Fabrication requires so much fixed capital that even treating them together—that is, as complementary stages in the overall production process—does little to escape the weight of dead labor; the combined ratio across the two only reaches about 8.75%.

There are a few important conclusions we can draw from this analysis. First, and most important, living labor is disproportionately employed in the design process over that of fabrication. It is more highly remunerated and contributes to a bifurcation along not only functional but geographic lines as well. This division works in numerous ways to further disadvantage racialized labor forces in non-Western countries, as Sean Cubitt argues more broadly:

> Technological advances in automation are aimed at reducing the living labor required to produce goods, and the falling rate of profit drives it to surprising levels of invention, driving down employment. However, even invention has costs, so when offshore manufacture and assembly provide cheaper labor, fewer taxes, and weaker health, safety, and environmental protection laws than in the Global North, the only reserved elements of industrialization to remain human dependent are marketing and design.[33]

The US government report supports this analysis, noting "the R&D and design stages . . . are capturing an increasing portion of the value of the semiconductor industry."[34] While they consider fabrication as "front-end production generally taking place in higher-income countries," with "assembly, testing, and packaging . . . often tak[ing] place in lower income countries," in both cases these are increasingly located in Asia, with the former in richer and the latter in poorer countries within the region.[35] In many ways, this is a harbinger of 100% utilization. Although Cubitt considers these "human dependent" elements susceptible to machinic replacement as well, in the semiconductor industry the cost of human design labor, especially for special-purpose processors beyond the CPU, is increasing. It is thus growing in precisely the area where automation is still comparatively light: the logical, rather than physical, construction of integrated circuits.

This is especially interesting since, as we have seen Friedrich Kittler suggest in the previous chapter, contemporary circuit design is a vastly computer-aided process. Nevertheless, computers for designing circuits are cheap relative to the equipment necessary to make them. Yet the accompanying labor, however assisted by digital mediation, has not (yet) been automatable to the same extent it has in manufacturing. Circuit design labor pushes forward the performance limits that general-purpose computing increasingly finds itself up against. It does so by developing bespoke schematics that increase computing effectiveness on specific tasks. It is profitable enough that capital has been willing to accept the higher wages of more developed countries where this expertise is (so far) centralized. The resurgence of the human in the face of stagnating computing power growth could therefore originate from this labor distribution within the semiconductor industry itself. Yet, at the same time, efforts to leverage AI in this area are ongoing and have already produced "superhuman" results, at least if press releases are to be taken at face value.[36] Regardless, such attempts are to be expected, given the drive to replace variable with fixed capital.

Correspondingly, another implication of the extreme imbalance between labor and automation across these two aspects of production is that vast economies of scale are operative in fabrication. In this sense, semiconductor fabs are a prime example of the dominance of machinery, which transfers rather than creates value, that Marx identified as a key aspect of the industrial age:

> Machinery, like every other component of constant capital, creates no new value, but yields up its own value to the product it serves to beget. In so far as the machine has value and, as a result, transfers value to the product, it forms an element in the value of the latter. Instead of being cheapened, the product is made dearer in proportion to the value of the machine. And it is crystal clear that machines and systems of machinery, large-scale industry's characteristic instruments of labour, are incomparably more loaded with value than the implements used in handicrafts and in manufacture.[37]

At the same time, industrial production resists the tendency of increasingly expensive apparatuses to drive up the cost of commodities through economies of scale. Here, "the greater the productive effectiveness of the machinery . . . the greater is the extent of its gratuitous service," as an increasingly minimal fraction of a machine's increasingly large value is transferred to each individual commodity.[38] In the case of integrated circuits specifically, even as the instruments of production increase in cost, decreasing component size allows their value to be spread ever more thinly as process size scales down. This is a direct economic benefit of the exponential diminishment in scale Moore's law promises, for as long it remains possible, economically or physically.

Transistor Size Versus Headcount

To sum up, despite ongoing efforts at automation, the semiconductor industry is faced with a growing portion of labor situated outside the manufacturing process itself yet deeply necessary to the production of newer commodities. At the same time, fabrication is increasingly dominated by machinery. There, increases in effectivity, largely tied to the scale at which this technology operates, drive decreases in the overall cost of manufacturing each integrated circuit component while at the same time dramatically increasing performance. To be cost effective, these reductions must balance out rises associated with acquiring and implementing new and improved equipment. We can now unpack the apparent contradiction between economic and physical limits in Moore's law.

Although coined after Moore's 1965 article, as Arnold Thackray, Brock, and Rachel Jones have argued, the origins of Moore's law can be traced to a lengthy chapter Moore published in 1963 as part of a collected volume covering the rapidly miniaturizing field of electronics. "In this hundred-page essay," they write, "Moore would achieve the seminal intellectual breakthrough of his life . . . demonstrating once and for all that the microchip was the future."[39] There, as they also note, Moore wrote that "the critical role played by the available technology in determining the applicability of microcircuitry in a given application cannot be overestimated. It determines what can be done at all, as well as what can be done economically."[40] The connection—but also distinction—between technology and economics can thus be found in arguably the earliest formulation of steadily increasing circuit density as a trend, and it is wrapped in the language of determinism. For Moore, the last instance was clear, and it was not economic.

At the same time, the gap between the state of the art in the early 1960s and any hard material limit down the road meant that Moore's law would play out in the socioeconomic as much as, if not more than, the realm of physical possibility. "To push forward the manufacturing technology," his biographers write, "engineers would have to design microchips that optimized cost, performance, and reliability, while system builders would have to form their products around those microchips. Gordon Moore's 'economics of integration' required a social framework."[41] Unlike many of his subsequent formulations, his *Microelectronics* chapter represents a rare instance in which labor factors explicitly, rather than implicitly. In addition to repeated invocations of the similarly etymologically derived "laboratory,"[42]

Moore describes, in some detail, the interrelationship between machinery ("technological capability") and scientific labor:

> The technological capability determines into which class the structure will fall—if it will be a manufacturable item suitable for design into operating equipment; if it will be a structure whose 'feasibility' can be established by the construction of a few samples by highly trained laboratory technicians; if it will be a one-of-a-kind structure made through the extensive labors of engineers adjusting and selecting until operation is finally achieved, suitable for description in a technical publication or contract report; or if it will fall into the class of *gedanken* structures, unmakable, but with principles of operation consistent with established physical laws.[43]

He further notes the potential "for production workers" to perform "well over 100" distinct operations while fabricating a wafer.[44] Projecting this number to grow, he writes, "Even so, it appears certain the increased processing complexity will more than justify the additional production steps by offering greater electronic function for a given cost."[45] Manufacturing labor, though less directly remarked upon here than that of design and experimentation, was plainly substantial.

Differences between the fabrication process at the time and the comparatively infinitesimal scales of the contemporary era notwithstanding, this initial analysis reflects a recurring labor–machinery dynamic under capitalism. Labor fills in where machinery is unable—here, to push the envelope of what it is possible for integrated circuitry to do, albeit at greater expense. In the early 1960s, the process of creating microelectronics was comparatively less automated than it is today. Yet with so many stages, Moore writes, manufacture "can function only if most of the individual steps have yields very close to 100%. These operations can be this well controlled."[46] Already present at the birth of the industry, the potential for further automation clearly persisted.[47] As a result, it may now be easier and more cost effective to design larger circuits, as Cerebras does, than rely on experimental, low-yield processes pushing beyond industry-standard scales. Either way, however, someone has to pay for "the extensive labors of engineers" to push the state of the art forward.

Yet any additional costs associated with these labors, in Moore's analysis, are dwarfed by the advantage delivered by shrinking components. While fabrication technology establishes what is possible, "the performance of microcircuitry is determined by the structural features employed for achieving the desired electrical functions. . . . In the final analysis, the structure itself is the important consideration, rather than the technology by which it is achieved."[48] By structure here, Moore means the composition of components—intricately linked to, but distinct from, how they are made. Structure, truly the last instance, is defined in large part by size such that, in integrated circuits, "costs of components are now primarily determined by the area they consume on the semiconductor wafer."[49] Consequently, "die size and yield are of prime importance in determining the economics of integration."[50] Moreover, "extra processing steps required to extend the available technology . . . are relatively inexpensive," and more than pay for themselves, Moore argues, if they drive further size reduction.[51] Far from being distinct from size, then, cost is intimately associated with it: Smaller components cost less.

Marx came to the same conclusion about industrial production in general, particularly as it relates to increasingly expensive machinery. "Given the difference between the value of the machinery and the value transferred by it in a day to the product, the extent to which this latter value makes the product dearer depends in the first instance upon the size of the product, on its area, so to speak."[52] The more commodities that can be physically processed by a machine before it breaks—before the second law of thermodynamics reduces it to an inoperable state—the less each individual commodity will contain of the machine's overall value. This dynamic is further modified by the state of wages at any given moment such that, "for the capitalist . . . the limit to his [*sic*] using a machine is . . . fixed by the difference between the value of the machine and the value of the labour-power replaced by it."[53] The cost of increasingly expensive machinery may thus go down, in relative terms, if wages themselves rise, as they are already doing in the semiconductor industry.

This is yet another reason to be skeptical of improvements to semiconductor density petering out in spiraling costs. Whether or not companies like ASML are able bring the *absolute* cost of their products down tomorrow, wage growth can have the same effect, making the price of their machinery *today* more appealing to capital in the future. However, the localization of payroll increases in the design rather than manufacturing side of the industry makes it possible for the latter firms to disavow this dynamic, at least for now. However, demands from fabless companies, their customers, to make up for the disproportionate labor costs they shoulder with even cheaper manufacturing could easily force them to face it. Here Intel's position as an IDM is highly revealing. On the one hand, it is attempting to leapfrog TSMC by acquiring the very lithography equipment its competitor has (for now) rejected.[54] On the other, the company announced substantial workforce reductions just a few months later, and more the following year.[55] Responsible for both design and fabrication, the company has chosen to invest in cutting-edge machinery and trim its workforce.

"Economic death" as an end to Moore's law thus appears plausible from a perspective immanent to capitalism. This is especially true given the hidden struggle between manufacturers and designers for the distribution of profits, a contest the latter are currently winning despite the proportionally higher wages they must pay. By contrast, the potential for an absolute halt to miniaturization in the form of physical limitations threatens to end any such arbitrage between these sectors, or indeed, between labor and machinery tout court. Rather than being distinct from cost, size plays such a large function in determining it that capitalist dynamics highly favor an exploration of miniaturization up to its material limits in the long run, even if the industry's commitment to doing so may falter in the short term. Viewed this way, any paradox is resolved. A point beyond which components cannot be physically shrunk represents a truly insurmountable cost barrier.

Perhaps this intertwining of the economic and the technological is an example of what Jameson means when he suggests that "Marx's idea of production," characterized by a firm distinction between the two, "has often been displaced and stereotypically tainted by its period association with that late nineteenth-century heavy industry today itself displaced by cybernetics and information technology."[56] Just as a classically Marxist understanding of political economy has provided extraordinary insights into all levels of human society and culture, so too can a technical understanding of the function and limitations of computing power enrich analyses of our own digital and incipiently post-digital era. One realization a theory of 100% utilization can offer is that both capitalism and digital technology rely on continual growth for their own perpetuation. It may ultimately be no coincidence, then, that the two should experience difficulties in continuing their prodigious levels of expansion simultaneously.[57]

High-Tech Slowdown

The interaction between these two poles does not play out only at the microeconomic level of the semiconductor industry and the shrinking dimensions at which its components are manufactured. It also unfolds across a macroeconomic realm characterized by an increasingly globalized and computerized mode of production. It is thus important to consider the relationship between political economy and information technology from this perspective as well. In doing so, we can begin to see what is at stake as their dynamism breaks down.

Approaching first from the side of capitalism and its discontents, computation's association with an immaterial finance capital, itself understood as secondary to physical, commodity production, has led some critics to doubt the impact of digitization. Despite the substantial amount of capital invested into chipmaking, Marxist geographer David Harvey, for instance, has argued that "information technology is the privileged technology of neoliberalism. It is far more useful for speculative activity and for maximizing the number of short-term market contracts than for improving production."[58] This attitude is reflected in at least one possible solution to what liberal economists have defined as the "productivity paradox," in which, as Harvey implies, the readily apparent impact of computerization does not seem to manifest itself in official growth reports.[59] In his classic article outlining this antinomy, Erik Brynjolfsson summarizes this "redistribution hypothesis": "IT rearranges the shares of the pie without making it any bigger. . . . For instance, IT may be used disproportionately for market research and marketing, activities which can be very beneficial to the firm while adding nothing to total output."[60] Harvey's overarching thesis, that "neoliberalization has not been very effective in revitalizing global capital accumulation, but it has succeeded remarkably well in restoring, or in some instances . . . creating, the power of an economic elite,"[61] is consistent with this premise.

A consideration of the digital platforms that have solidified their dominance over the last decade further supports this theory. At the same time, however, it also makes the case for computing as a central, rather than peripheral, aspect of contemporary capitalism. Firms like Facebook, Amazon, and Uber, accelerationist theorist Nick Srnicek has argued, arise out of a low-profit neoliberal environment caused by the overproduction of goods and factories in the wake of the Second World War. Capital thus sought valorization in other ways, most significantly through the deregulation of the internet in the 1990s and then via subsequent financial bubbles culminating in the crisis of 2008.[62] This series of booms and busts consolidated a wave of businesses "centred upon extracting and using a particular kind of raw material: data."[63] Digital platforms are thus incentivized to do more and more to keep users active, providing this information. As data became a privileged source of value in a post–Lehman Brothers economy, he contends, physical products increasingly became services. Under this paradigm, even manufacturers have begun to compete by leveraging the information they collect in order to solidify and expand their market share, independent of financial and productivity growth.[64] A redistributive platformism comports with a view of computation—like capitalism and especially in its globalizing, financial stage—as a totalizing force, even were it the case that IT had *zero* effect on productivity, something even most skeptics do not concede.

Srnicek sees similar all-encompassing tendencies in platforms themselves, whose "ability to rapidly scale . . . by relying on pre-existing infrastructure and cheap marginal costs means that there are few natural limits to growth."[65] True to his theoretical investments, however, this characterization of platforms as nearly unbounded does not take the material exigencies of computation into account. As he argues, such businesses are uniquely enabled by "a low-interest rate environment. . . . Rather than a finance boom or a housing boom, surplus capital today appears to be building a technology boom."[66] However, we should not understand from this, as he sometimes appears to, that computing and finance are conterminous. The era of cheap money described has come to an end as central banks seek to tame the inflation resulting from an overdetermined economic situation comprised of pandemic-era supply shocks, postvaccination employment recovery, war, price gouging, and more. In this new fiscal environment, we will discover the real worth of the data platforms collect. However, both in their origins and to whatever extent they end up persisting, these companies are deeply reliant not only on financing but Moore's law. Their business model was initially enabled by price reductions in "the technology needed for turning simple activities into recorded data,"[67] and its survival hinges on it. "If these platforms wish to remain competitive," Srnicek writes, "they must intensify their extraction, analysis, and control of data—and they must invest in the fixed capital to do so."[68] Increasingly powerful forces of production are thus as essential as easy access to the capital that funds them. Moreover, as any capitalist knows, these two things are not identical.

Here, then, the parallel between computing power and what Marx called the "intensification of labour" is at its strongest, with the caveat that, unlike workers, machinic intensification does not need to take place within the confines of a shortened workday.[69] It is the specific innovation of platforms to extend their data collection and valorization around the clock via sleep trackers, "smart" appliances, always-on cloud services, and the like. As Jonathan Beller writes, "Networked relations have colonized the social field, created 'digitality' and the 'social factory,' and maximally extended the working day to create a near total occupation of life."[70] This has led to a paradoxical situation in which, from the perspective of our devices and the corporations using them to extract data, their "workday" can in practice be treated as a limiting factor precisely because it is already growing to include every moment of our existence.[71] Every second of each twenty-four-hour period can only be segmented further to extract more out of it. This is precisely the goal of increases in clock speed and now parallelization—of computing performance in general. Technologies enabling the capture and processing of data must be continually improved. Any risk of 100% utilization thus presents a threat to not only the chip industry, but further market consolidation and temporal expansion.

At the same time, and just as it is with industrial machinery, both this drive toward intensification and the limits it must eventually encounter have been and will be dispositive not only for platforms but *any* attempt to extract value out of digital mediation. For example, there is every indication that finance capital is itself acutely concerned with the material foundations of computing. As Edward Nik-Khah and Philip Mirowski argue, for a neoliberal economics that explicitly understands it as an information processor, "the market is a person–machine system, with the thinking offloaded onto things."[72] The increasingly complex calculations and network operations that follow from this, for example in high-frequency trading, thus demand ever-greater speed.[73] "Financial derivatives and digital media platforms" alike, for Beller, "ultimately require the movement of the exponentially more complex pathways and calculations of capital into the (nano)second-by-(nano)second operations of discrete state machines," inaugurating "a new phase of the capitalist mode of production, the computational."[74] Marc Steinberg, meanwhile, has posited that the digital nature of securities transactions and tech stocks' dominance in the market suggest the relationship between finance and platforms could run opposite to Srnicek's characterization. The former, he writes, "may itself be a subset" of the latter.[75] It is more accurate, however, to understand both as increasingly reliant on computation *in general*. In this sense, processing power finances finance and platforms platforms. Following the NRC report, many other industries can be seen as dependent on computing speed as well.

The productivity paradox may at first blush cause us to think otherwise, however. It could lead us to assume, like Harvey, that computing is only or primarily relevant to semiotic, superstructural aspects of society and not manufacturing itself. Like all contradictions, though, the productivity paradox needs inhabiting rather than dismissal in favor of either of its extremes. Its unresolvable status is simultaneously a reflection of the reality we inhabit and our presently inadequate rationalization of it. This is further borne out by the productivity paradox's complicated history. While it is now invoked to suggest that commodity production does not overly benefit from computerized automation, it was previously used to illustrate how *services* were falling behind technological gains in manufacturing.[76] Subsequently considered to have been reconciled in the 1990s as productivity caught up with IT expenditure, more recent work, in challenging this conclusion, has at the same time highlighted one area where the correlation between computerization and productivity over this timeframe held: the manufacture of information-technological goods themselves.[77] This effect, which attenuated around the turn of the millennium, just before the raw clock speed of processors themselves began to stagnate, nevertheless supports the notion, alluded to earlier, that the last decade of the twentieth century was a critically generative moment for digital mediation.[78]

The multimedia boom of the 1990s laid the groundwork for the absorption of analog media into the digital machine, a key moment in computational totalization.[79] This capability is a direct reflection of increases in computing power.[80] As Harvey goes on to note, "The main arenas of production that gained [from computerization] were the emergent cultural industries (films, videos, video games, music, advertising, art shows), which use IT as a basis for innovation and the marketing of new products."[81] Technological developments at the end of the twentieth century thus align with the redistributive thesis even while going beyond it, suggesting that growing productivity in electronics manufacturing at the time may have cannibalized other areas of commodity production. As Jonathan Sterne argues in his analysis of how the MP3, as a binary file type, has contributed to the disappearance of physical formats (including earlier digitally encoded artifacts like the compact disc), these developments effectively shifted revenue from record companies to technology manufacturers (or from the former branches of multinationals with interests in both to the latter).[82] This process, he notes, following music critic Philip Sherburne, is not in fact immaterial but rather one in which physicality persists, only at microscopic scale.[83] Rising to prominence over the 1990s and early 2000s, such miniaturized, "universal" storage and media products, capable of working with any stream of digital information, are increasingly made up of little other than silicon chips. They therefore face a limit in terms of both storage capacity and its trade-off with encoding and decoding power should semiconductor density level off.[84]

Over the past thirty years, these shrinking digital devices have indeed replaced an ever-widening array of physical media.[85] Computerization's impact on production should therefore be understood in relationship not only to financialization and *increased* productivity (or the relative lack thereof) but *diminished* commodity manufacturing, the disappearance of things whose functions have become replicable in digital logic. Consequently, the question of how far this logic can proceed must be understood in terms of the limits to computing power. Again, the extent of digitization's reach, whether in a platform/data collection capacity or otherwise, depends on how much of the broader physical world its operations can potentially replace. This, in turn, depends on the complexity of the material processes in question and therefore the capacity of computing power to simulate them.

Brynjolfsson places the redistributive thesis on the opposite side of a divide between ontological and epistemological solutions to the productivity paradox from that of his preferred answer, that this contradiction is simply a case of "mismeasurement."[86] However, the concept of digitization subsuming some portion of previously analog activity and the idea that productivity gains are not captured by our existing understanding are hardly incompatible. Some as-yet ineffable quality could both result in an advantage for companies or sectors that do adopt information technologies *and* not show up in the data. Brynjolfsson's recent work with Avinash Collis attempts to quantify this precisely in terms of negative value, or "consumer surplus," by asking people what remuneration they would accept to forego a free digital service they previously would have had to pay for in physical form.[87] Despite the ongoing contradiction presented by the productivity paradox, and the way the proliferation of lost (exchange) value he and Collis chart suggests a limit to commodity relations, Brynjolfsson has come to be known as one of the leading proponents of the idea that automation will rapidly upend the workforce—without, of course, overturning the relations of production themselves.[88] As environmental scholar Anne Pasek has suggested, however, in one of her analyses of Moore's law, "getting closer to the material foundations of digital operations might, paradoxically, help us better apprehend and critique the seeming dematerialization of digital media."[89] A potential limit to computing power in essence forecloses the absorption of more complex physical processes. It thus places a boundary on the progress of informationalization we have seen over the lifetime of Moore's law, and especially since the 1990s.

This is the view from the side of political economy. From the side of computation, the situation looks even more extreme. From its inception, computer science has placed an emphasis on the simulation of real-world processes, beginning with labor. Turing's description of a universal machine, the conceptual basis of all classical computers, is directly predicated on an analogy with a human working out a calculation.[90] While possible to conceive of such a machine taking input from people, he writes, the one his foundational paper proposes is "automatic," in the sense that "at each stage the motion of a machine . . . is *completely determined* by the configuration," and "universal," such that "a single machine . . . can be used to compute any computable sequence."[91] Turing not only inaugurates a revolution in mathematical materialism here, as Paul Cockshott has argued, he presents a mechanical procedure as a replacement for a certain kind of *labor*. In doing so, he further demonstrates that type of labor an effective mediator for any calculable operation. Anything that can be quantified can be automatically computed, at least up to the logical limits immanent to that form of rationalization—the true object of his proof.[92] Inside these epistemological boundaries, as subsequent decades have demonstrated, the operating question is not *what* can be done, but *how fast*.

This focus on speed places practical computing well within the same framework of capitalist production that demands continual reductions in integrated circuit component size. Mid-century contemporaries like cybernetician Norbert Wiener quickly seized on the connections between the formalization of procedure Turing proposed and the quantification of factory labor in the early twentieth century. In contrast to Vannevar Bush's analog Differential Analyzer, Wiener wrote, a machine for solving *partial* differential equations

> would have to work at extremely high speed, which to my mind put mechanical processes out of the question and threw us back on electronic processes . . . in addition to including an arithmetical mechanism, it would need a logical mechanism as well and would have to be able to handle problems of programming on a purely logical and automatic basis. The notion of programming in the factory had already become familiar through the work of Taylor and the Gilbreths on time study, and was ready to be transferred to the machine. This offered considerable difficulty of detail, but no great difficulty of principle. I was thus convinced as far back as 1940 that the automatic factory was on the horizon.[93]

Programming, at least in Wiener's accounting, entered the computational lexicon through the factory.[94] What were merely configuration tables for Turing became Taylorized instructions. As in the manufacture of commodities, "the speeding up of the pace of a machine leads to a directly proportional increase in the number of operations which can be performed in a given time."[95] As we will see in chapter 6, this understanding of an essential interchangeability between factory and computer is furthered by Wiener's extension of thermodynamic principles from their industrial roots in steam engines to mid-century cybernetics and computing. Presently, however, we can instead observe how automation, speed, and economies of scale have been both the premise and object of much foundational thinking on the development of the discrete, electronic computer.

At the same time, the promise of computation goes further than this, holding out the hope, as we have seen in the introduction, that the entire physical world can be simulated or reproduced via its methods of rationalization. If not only factories but the whole universe could be accurately replicated inside a computer, then the productivity paradox (among many other things) would be essentially solved. It is telling that this type of thinking appears in both the originator of our modern theory of computation, Turing, and the man who, concurrently and without awareness of this development, put it into practice: Konrad Zuse. As Turing famously wrote, "There is a remarkably close parallel between the problems of the physicist and those of the cryptographer," though he was wise enough to extricate himself at the last minute, adding that "the correspondence is very close, but the subject matter of cryptography is very easily dealt with by discrete machinery, physics not so easily."[96] Crucially for our purposes, Zuse expands upon this difficulty by pointing out that the relationship between computer science ("data processing") and the formal rationalization of physical processes is dialectical. As programmers come up with practical implementations of mathematicians' models, the latter often adjust these models in light of what is technically feasible.[97]

The threat of a limit to computing power therefore not only raises the question of how far this desire to simulate can be pursued, but also what it may have left out in service of this mission. As the starting point of computational simulation, labor is the ideal place to center an exploration of this issue. While it is the remainder I am primarily concerned with, it is not the only one. We are obviously nowhere close to being able to simulate the universe yet, if such a thing is even possible. But we have to acknowledge these ambitions lie at the foundation of modern computing, which, like capitalism, begins by appropriating labor and ends with the quantification of everything in sight.

Across financialization, "dematerialization," and physical processes including (but not limited to) manufacturing, then, computer power has multiplied the reach of the industrial dialectic between people and machines into ever greater realms of existence. In many ways it makes sense to consider each of these categories a type of "automation": the increasingly automatic transfer of value out of commodity production and into high-frequency trades and digital pseudo-objects on the one hand and the perhaps slower-paced but undeniably real extension of computerized tools, methods, and platforms into the factory as such (a massive electromagnetic pulse in Shenzhen or the maquiladora corridor would have substantial economic effects) on the other. Labor, then, is being replaced—in more ways than one—by a proliferation of microchips that has long since surpassed the number of human beings on the planet. At the same time, the capacity of future generations of integrated circuits to perpetuate the rate of expansion computerization has delivered for nearly a lifetime now is increasingly in doubt. In such a conjuncture, the unique capabilities of labor—above and beyond those that can be replicated or abstracted away via cutting-edge machinery—shine through.

A physical analysis of computation, then, has something to tell us not only about the parameters of digital mediation but those of political economy as well. This is especially true because, as Srnicek points out,

> the digital economy is far more important than a simple sectoral analysis might suggest. In the first place, it appears to be the most dynamic sector of the contemporary economy—an area from which constant innovation is purportedly emerging and that seems to be guiding economic growth forward. The digital

> economy appears to be a leading light in an otherwise rather stagnant economic context. Secondly, digital technology is becoming systematically important, much in the same way as finance. As the digital economy is an increasingly pervasive infrastructure for the contemporary economy, its collapse would be economically devastating. Lastly, because of its dynamism, the digital economy is presented as an ideal that can legitimate contemporary capitalism more broadly.[98]

This statement reaffirms some of the points made here and in the National Research Council report, while arguably contradicting Srnicek's characterization elsewhere of tech's relationship to finance. I contend that the systemic centrality of digitization across separate areas of economic activity requires its own reckoning alongside, rather than subordinate to, finance capital. So, too, does the extent to which it serves today as a figure for economic development writ large, and thus a justification for the existing relations of production, up to and including the ways it works to obscure the centrality of labor. A route back to the profound, universalist insights of Marxist materialism must therefore be found by way of the operations of those devices most closely approximating Turing's so-called universal machine, apparatuses which have ostensibly been successful in *de*materializing so many forms of work and media over the past fifty years. The end of Moore's law could be disastrous for a political economy that requires the ever-intensifying powers of technology to stave off full employment. By the same token, however, it may be more of a boon for efforts to recenter our globally networked society on human capabilities and needs.

Yet, instead of trying to cruelly ration the existing surplus of computing power as it does the reserve army of labor, a digitally dependent capitalism is moving full speed ahead (inasmuch as such a thing continues to be possible) with extracting the remnants of this computational reservoir from existing paradigms. As we have already begun to see, this push has given rise to new hardware designs eschewing the general-purpose CPU in favor of more specialized configurations. It has also driven corresponding changes in software engineering. Such circumstances render the concept of computation increasingly unstable and precarious. In some ways the fate of ongoing digitization now parallels that of workers in a casualized labor force. In others it follows Turing in assuming responsibility for more of their attributes. This is especially so for those related less to manual than cognitive labor—illustrating the validity of concerns over the rise of artificial intelligence, if perhaps not the most extreme projections. These developments are the subject of the following two chapters.

3 Squeezing Silicon: From Software Control to Hardware Efficiency

Increasing constraints on the perpetuation of Moore's law have foregrounded the contradictory relation between hardware, as a configuration of tangible matter, and software, as a form of symbol manipulation. The growing tension between these two aspects of digital mediation has profoundly affected not only labor dynamics in the semiconductor industry but the nature and function of circuitry and programming more broadly. In particular, the recent focus on field-programmable gate arrays (FPGAs) and application-specific integrated circuits (ASICs) highlights a drive to make hardware more software-like and vice versa. FPGAs enable the reconfiguration of circuit layouts postfabrication, while ASICs are chips hard-wired to better perform the necessary calculations for a particular task. In both cases, these devices are more suited to some endeavors than the CPUs typically found at the heart of classical computing. FPGAs are especially attractive, for example, to retrogaming enthusiasts seeking to accurately run software for systems that are no longer manufactured. ASICs, on the other hand, are perhaps most (in)famous for their applications in cryptocurrency mining where, despite their efficiency, they draw a tremendous amount of energy. This chapter will therefore follow the logic of computational innovation "down" from the software-based, symbolic user interfaces that form our everyday phenomenological experience of personal computing into these changing physical configurations of silicon, and then back "up" into the aggregated collections of hardware, cryptocurrency miners and otherwise, in data centers. Ultimately, by pushing the rationale of continual improvements in processing power to its extremes, we can more clearly understand the implications 100% utilization holds for the point where classical and postindustrial capitalism most overlap: the efficient use of energy.

Critically, this material customization of hardware for specialized purposes is not proof of software's ultimate jurisdiction. Rather, as we have already begun to see, it is *determined* by encroaching conditions of 100% utilization. As Mark Pesce puts it, with the pace of the more comprehensive form of computation represented by CPUs slowing, hardware becomes more particular and less "universal" to make up for it. "That push toward 'general purpose computing' gave us thirty years of Wintel," he writes, "but that no longer looks to be the main game. The CPU is all grown up."[1] As custom and customizable chips demonstrate, this shift in our computational foundation goes beyond either refactoring software to take advantage of multiple processors or expanding the raw area taken up by individual chips. It extends into material modifications to silicon-based computation itself. Taken together, these technologies' drive toward increased efficiency outside of the strictures of Moore's law emphasizes the dialectical interdependence, and incipient exhaustion, of conventional hardware and software paradigms alike.

As we have already seen, Turing machines are at once a material instantiation of computation and the introduction, alongside it, of a semiotic component. Indeed, the two are, in some profound sense, inseparable. Turing's concept of configuration tables introduces and enables a far greater one: that of a programmable—and thus "universal"—machine.[2] A computer without the capacity for configuration would be a *particular* machine, a tangible manifestation of a specific procedure for solving a given problem rather than computable problems as a whole; conversely, an algorithm lacking a material basis would be little more than a concept.[3] Viewed this way, we are once again thrown, at the heart of computation, into philosophical tensions that also lurk behind the broader sense of intensifying stagnation accompanying our moment of exhausted political-economic expansion: between the universal and the particular, the global and the local, the abstract and the concrete, and the ideal and the material.

To view these agonisms as dialectical, then, is not only to incorporate a sense of their potential sublation, of the supersession of such oppositions. It is also and in advance of this final moment an understanding, following Fredric Jameson's crucial analysis, of their capacity to shift valences, to privilege one pole of this dyad under one set of circumstances, its correlative under another.[4] The software–hardware dialectic in this sense befits our moribund era, one in which all available solutions to a given problem (as with the binaries enumerated above) frequently come across as equally problematic. Within this opposition we can find matching, and comparably forceful, claims on both sides of the divide it represents. As we have already seen, Friedrich Kittler's essay "There Is No Software" has become something of a strawman for the position privileging hardware. Instead of engaging with the full breadth of its argument, namely, that software as a concept is contingent upon a material instantiation of discrete, electronic, silicon-based Turing-like machines conditioned by language and economics—"letters and coins . . . books and bucks"—that offers a limited range of what computation is and could be, even well-informed commentators have relied on its title as a shorthand for the position that software is irrelevant *today*, under present conditions.[5]

On the opposite side of this equation and, as Jussi Parikka and Erkki Huhtamo point out, more in line with an Anglophone insistence upon the primacy of symbolic factors often opposed to the materialist concerns of either Marxism or German media theory, we can situate Lev Manovich's book *Software Takes Command*.[6] Named in homage to Siegfried Giedion's study of the rise of mechanization, the difference in titles recapitulates the shift from industrial modernity to digital postmodernity, and with it one in emphasis from the physical to the conceptual–representational.[7] This transition, for Manovich, inaugurates a "world of permanent change—the world that is now defined not by heavy industrial machines that change infrequently, but by software that is always in flux. . . . Software has replaced a diverse array of physical, mechanical, and electronic technologies."[8] Software, here, measures the progress of dematerialization and is the technical manifestation of a neoliberal imperative to "embrace change."

Rightfully identifying the late 1980s and early 1990s as a turning point for digital media, Manovich argues, directly against Kittler's "classical modernist move," that "while the progress in hardware and Moore's Law played crucial roles in this development, even more crucial was the release of software with a Graphical User Interface."[9] However, toward the end of his book Manovich reveals he knows very well the role and even dialectical properties of hardware in his analysis of software's ascendancy. "This evolution," he notes, "is intrinsically connected with the parallel development of *hardware*. . . . For example, a computer display with 35,840 x 8,000 pixel resolution is not only quantitatively but also qualitatively different from a display which only has 1024 x 768 pixels."[10] But all the same, here software not only exists, it is in the driver's seat. One would have to look closely through a list of linguistic factors to find, in the middle, the sole material determinant behind "the permanent extendibility of the computer metamedium" and the permanent revolution in knowledge it supposedly enables: "constantly increasing speed of processors and decreasing cost of storage."[11]

I have already emphasized the way in which Alan Kay's idea for a GUI-supported object-oriented programming environment was informed by his observations about the increasingly powerful computational capabilities that would unfold as hardware improved. Manovich takes the opposite tack, citing him as a "key protagonist" of *Software Takes Command*:

> When Kay's colleagues implemented various media editors on their 'interim Dynabook' in the first part of the 1970s, most of this software could not compete with their physical and electronic equivalent tools. . . . So, in some sense, the first period of media computerization—between the completion of Sketchpad in 1963 and the release of PageMaker in 1985—was theoretical. During this period, the conceptual principles and the key algorithms necessary for detailed simulation of physical media were developed, before the sufficiently cheap hardware was available.[12]

But, as we also saw previously, this "theoretical" vision was *itself* enabled by Moore's material analysis. Retelling its manifestation in the early 1990s, Kay even includes an image of Moore's famous logarithmic graph.[13] This recollection matches with the appeal to Moore's law, albeit not by name, in Kay's 1972 description of the Dynabook. There, he notes that

> just as in the HP-35 pocket electronic "slide rule", the main savior of our dreams is cheap LSI [Large-Scale Integration] components. The HP-35 uses five LSI chips with an equivalent of 30,000 transistors, for an average density of 6000 transistors/chip. Even better packaging densities are now being accomplished. The price of a packaged LSI chip seems to approach $12.00 asymtotically [*sic*] over a two-year period, then may dip suddenly to about $5.00.[14]

Kay's dissertation, written three years earlier, similarly argues that LSI memory "is the wave of the future, and there is a good chance that prices will come down to the .1 cent per bit level, where they should be."[15] It thus does not make sense to treat hardware and software developments as separate—even if complementary—factors in the transformation of the personal computer into a multimedia device. If not (yet) a sublation, then at least some sort of reconciliation is in order.

On the one hand, observations of the relative ease with which electronic circuitry could be shrunk and enhanced gave rise to increasingly expansive visions of what computers could do. On the other, increasingly affordable computing power made possible the mainstream availability of graphical interfaces by the mid-1980s. The ease of use and greater audiovisual capabilities these systems offered then drove adoption and, eventually, cultural change. Kay's mediatic innovations (like Engelbart's) were galvanized, in other words, by the realistic projection of sufficiently advanced hardware to support them and then realized by this assumption's material manifestation. Indeed, the popularization of user-friendly interfaces is generally one of the last steps in the maturation of a technology. We need only look at the astronomically high prices of pioneering graphical computers like the Xerox Star and Apple Lisa—both of which can trace their paradigms back to the experimental Xerox Alto worked on by Kay and others at the company's Palo Alto Research Center—to get a sense of the process by which such interfaces entered mainstream computing.

While development costs for these machines were obviously spread out between software and hardware, the Lisa in particular, though significantly more affordable than the similarly inspired Star, offered an especially disadvantageous combination of high price and low performance that further affirms the significance of computing power for the project of graphical interfaces and all that flowed from them.[16] As Kay, a pioneer of GUI computing with stints at both companies, observed, "the Lisa failed because it was very under-powered, and so while it did beautiful things, it did them very slowly," leaving the arrival of the GUI into mainstream personal computing to the introduction of the Macintosh the following year.[17] In other words, it was *speed*, rather than software, that doomed the first attempt at a mass-marketed graphical computer. If anything, the Lisa's software was *too* comprehensive, leaving little room for outside programmers to create their own products for the platform.[18]

By contrast, as we approach 100% utilization we can look to the state of the GUI to see that real innovation in this field barely exists anymore. Graphical desktop interfaces have largely converged around a set of principles that draw heavily from the PARC-inspired Lisa/Macintosh design while at the same time eschewing many of the rigorous stances of Apple's 1980s interface team. Barring augmented and virtual reality approaches that have remained, despite massive technological change, in an experimental state since their introduction in the early 1990s, subsequent UI innovation has mostly been concentrated around the newer set of affordances offered by the smartphone. There, however, programmers have responded to its walled-garden ecosystems and awkward tactility seemingly by embracing the most sinister implications of the scholarly analysis of GUIs as ideological. Smartphone apps routinely dupe users into sharing the personal information developers' employers deem profitable by hiding not only the fact that they collect it, but the settings allowing users to opt out, a trend that has subsequently expanded as far as desktop operating systems.[19] It is in this way that we arrive at the historical impasse, and paradox, confronting us.

The first step toward resolving it, as I have tried to take briefly earlier, is to historicize the rise of graphical software and the wave of digital multimedia it enabled. I am in full agreement with Manovich about both the significance of the late 1980s and 1990s in this regard and the valence placed upon software in particular and the semiotic in general during this period.[20] Yet, as historicization is wont to do, such an analysis reveals the material determinants behind the conceptual and representational elements Manovich privileges. Crucially, it also introduces the question of valence into the hardware–software dialectic. Where, under conditions of computational growth, this dialectic paradoxically favored software, as Moore's law has begun to peter out it demands a reevaluation of the physical basis of computation—that is, of hardware.

However it is not ideological mystification alone that accounts for the paradoxical situation in which we find ourselves. Since the late 1980s, software really *does* seem to have taken command. Moreover, it really does seem to have done so not only over the decade and a half when Moore's law was in full effect, but *even as* hardware performance growth has slowed. This is well and truly a contradiction if material improvements drive software development. It can only be reconciled by taking into account the other aspect of dialectics invoked via Jameson's reading at the outset of this chapter, namely that supposed opposites arrayed in this manner are not in reality opposites at all but form, in some unrealized sense, an identity. The slowest, least functional Turing-complete device is nevertheless technically capable of running the most complex software imaginable,[21] while, contrariwise, future hardware could physically instantiate procedures we typically enumerate today via code. The end of Moore's law could be said to place a *greater* emphasis on software, since the hardware platforms upon which applications have been traditionally built are in the process of stabilizing. However, it does so in surprising ways, by reshaping hardware in software's image. As Pesce writes, "The cheap and easy gains of the last fifty years of Moore's Law gave birth to a global technology industry. The next little while—somewhere between twenty and fifty years out—will be *dominated by a transition from software into hardware,* a confusion of the two so complete it will literally become impossible to know where the boundary between the two lies."[22] The paradox of 100% utilization, then, is that *Software Takes Command* as "Zombie Moore's Law Shows Hardware Eating Software."

The rise of digital media alongside exponential returns from Moore's law enabled expansive semiotic paradigms such as object-oriented programming and the GUI. The tightening of the platform's material constraints, in turn, has encouraged a shift back toward the physical realities of computation. As we saw in chapter 1, the first step in this change of valence was itself symbolically driven. The move to parallelization signified a return to the more explicit tailoring of software to hardware that defined computing in its early days and even much of the personal computer market until the late 1980s, as developers sought to optimally utilize limited hardware by writing in low-level assembly language.[23] The obverse step is to customize hardware for operations left previously to software applications, a strategy M. Mitchell Waldrop notes as an explicit goal of the semiconductor industry's transition away from Moore's law.[24] Yet, this, too, is in many ways a return, at a higher level of order and sophistication, to the principles undergirding integrated circuitry in its early days.

Contemporary electrical engineering approaches this problem from two opposing directions: the field-programmable gate array and the application-specific integrated circuit. Where the former allows software to reconfigure material hardware arrangements, the latter fixes routines that have typically been written in programming languages directly into the physical design of the circuitry itself. This guarantees faster performance by eliminating more of the abstraction layers between concept and execution. These dual approaches take the place, within hardware, of the dialectic between software and hardware. Where this latter dyad is in some sense a product of an era of rapid growth in computing power, the present juxtaposition of FPGA and ASIC solutions is an outcome of the historical conditions of 100%

utilization. It arises as engineers and developers try to make the most of existing hardware density now that an end to "inevitable" progress appears on the horizon. As such, these types of chips raise important questions about both the efficacy of computing and, more broadly, what all this processing effort is being used *for*.

FPGAs: Postponing the Inescapable

As their name suggests, field-programmable gate arrays extend the software ecosystem down into the connections between chunks of physical computing power, groupings of logic gates capable of realignment "in the field." FPGAs have proven especially popular in areas where chips that are no longer manufactured need to be recreated. Projects like the MiSTer retrogaming console enable vintage games to be played via the ad hoc physical configuration of compute modules for each emulated system rather than the typical software-based route. Descriptions of this process frequently not only emphasize the performance advantage such an approach delivers but make claims of increased verisimilitude in relation to the systems being emulated:

> The MiSTer platform runs a separate "hardware emulation core" for each console, computer or arcade board you run. In order to accurately create those cores, you essentially need to recreate the original hardware in HDL (Hardware Description Language), which is more similar to a schematic than traditional coding.
>
> To do this, you map out each component on the motherboard, trace all the lines and analyze how they "talk" to each other in real time. . . . Depending on the complexity of the board and platform, this total process can take hundreds, if not thousands of hours to complete, but if done correctly, it can result in a cycle-accurate FPGA core. These hardware "cores" are then flashed to an FPGA . . . whose logic elements are then physically rearranged to match.[25]

The HDL, usually a language like Verilog or its more complex derivative, SystemVerilog, reflects the mappings specified in hardware blueprints, rather than the higher-level abstractions we typically think of as code. These symbolic representations can then be compiled and uploaded to a FPGA, rerouting its internal mechanisms correspondingly. It is useful in this sense to compare FPGA programming with software on single-tasking systems, especially that written directly in assembly as alluded to previously. With such programs, a specific application will more or less take over responsibility for the operation of the entire computer. HDLs serve a similar purpose but at the level of physical configuration (albeit toward a particular logical aim).

Like integrated circuits themselves, HDLs allow for specific operations to be carried out either discretely or via analog means, reproducing the range of silicon transistor functionality. As such, they allow control over hardware performance at a much more granular scale than high-level programming languages. Similarly, they not only support both sequential and parallel procedures, but privilege the latter over the former, making them especially attractive in a computational environment characterized by stagnant clock speed. As Arduino, well-known manufacturers of open microcontrollers used in "maker" projects, introductory electronics courses, and so-called physical media arts curricula (and thus themselves an example of the contemporary return to hardware), put it in introducing their FPGA platform, "all hardware description languages have the concept of parallel statements which means that as opposed to software programming languages which execute instructions sequentially, here all instructions are executed at the same time."[26] As they go on to describe, sequential operations can be achieved in SystemVerilog by building discrete-state systems on top of this parallel foundation, as well as through expedient conventions in the way HDLs are interpreted by FPGA compilers.[27] This, then, is a straightforward example of software taking command, extending to a level of hardware below even that of machine code to condition performance-enhancing physical features, notably including parallelism.[28]

As the existence of the MiSTer suggests, FPGAs have garnered significant attention from retrocomputing enthusiasts and preservationists for their perceived advantage in accuracy. Yet it is crucial to insist that what is often framed as higher fidelity is in fact simply a matter of performance. Moreover, when taking place within media-theoretical circles, considerations of these hardware–software hybrids' precision can recapitulate earlier debates over the nature of code. Fabian Offert, for example, argues that FPGAs could aid archival and archaeological efforts to maintain access to obsolete and/or anomalous platforms (and the software designed to run on them) because such systems "contain a semantic surplus of circuit-level functionality, an additional level of meaning that exclusively rests on a number of low-level peculiarities."[29] Yet, paradoxically, this overflow is recuperated by its translation from hardware into software, specifically HDLs: "This makes it possible not to emulate, but to clone nearly every piece of computing hardware imaginable: *the FPGA becomes the hardware it is programmed to be.*"[30]

For Offert, this is a natural consequence of the way "that the distinction between software and hardware could be regarded as somewhat artificial, up to the point where we could say, with Kittler, that it is completely ideological."[31] But for Kittler it is *software* that stands in for ideology: a symbolic phantasm limiting, and ultimately superfluous to, hardware as a physical machine. It is not the case, in other words, that this distinction collapses into an undifferentiated mass called "computation." Rather, computation as a material procedure can operate with or without software directing it.[32] It is thus possible to complicate this view of FPGAs on both grounds, philosophical and practical, motivating Offert's proposals.

Conceptually, the notion that an "FPGA becomes the hardware it is programmed to be" reiterates Alexander Galloway's assertion about programming languages, now seemingly transposed from software into hardware. It is in fact the *same* claim—not because software and hardware are fundamentally indistinguishable, but rather because *Galloway's initial claim is already situated at the interface between hardware and software*. Let us revisit this moment from the introduction with the addition of slightly more context: "Code is the first language that actually does what it says—it is a machine for converting meaning into action."[33] "Action" here, can only be understood as material effectivity, as making the computer *do something*. Since its initial formulation, this has been understood in technical terms as a change of state, the Turing machine's particular configuration and position vis-à-vis its input tape. These are a result of its previous or initial state, corresponding location along that tape, and processing of the information thereon. Further developments we would be more likely to subjectively consider "actions," such as the turning on of a light, the (re)presentation of a sound, image, or message, and so on, follow from these internal transformations. A change of state could, for example, close a circuit or series of circuits resulting in the flow of electricity powering a bulb or pixel.

Programs written at levels higher than that of an HDL, in other words, *already* place universal machines into specific configurations, even though they do not physically reconfigure circuit layouts to do so. It is what machine language itself, to which all higher-level programs and programming languages compile down, does to hardware: temporarily direct a general-purpose computer's material operation in a distinct way. As the late, well-known emulator developer Near, writing under an earlier pseudonym in a since-deleted blog post, put it, a HDL "tells the component how to operate, in the same way that C++ instructs a general purpose CPU how to operate."[34] While hardware description languages allow for continuing this arrangement beyond (some of) the specific material limitations of a given hardware configuration, they effectively do so by extending the logic of code into hardware itself. Consequently, all the torturous paths between description and execution outlined by Chun in response to Galloway also apply to FPGAs, which, in the case of retrocomputing applications like the MiSTer, routinely receive updates to their "cores" or hardware description files. This would, of course, be unnecessary if FPGAs had the same ontological status as the hardware they emulate.[35]

This does not mean, however, that there are not concrete distinctions between software and hardware when it comes to their effectivity in the physical world. It is in this, more practical respect that FPGAs offer an advantage over general-purpose hardware and higher-level software language implementations. Again, the world of emulation here is illustrative. As Near argued at greater length in an article for the technology website *Ars Technica*, commodity PCs, given they are sufficiently powerful, can also achieve "cycle-accurate" emulation:

> It doesn't take much raw power to play Nintendo or SNES games on a modern PC; emulators could do it in the 1990s with a mere 25MHz of processing power. But emulating those old consoles accurately—well, that's another challenge entirely; accurate emulators may need up to 3GHz of power to faithfully recreate aging tech. . . .
>
> But this accuracy comes at a serious cost. Making an emulator twice as accurate will make it roughly twice as slow; double that accuracy again and you're now four times slower.[36]

Computing power, then, ultimately determines the accuracy of emulation. It is the fundamental barrier to any recreation of a hardware platform on top of a different material foundation, even those that operate much faster. Indeed, the same kind of decapsulation, or "decapping," process of discontinued chips that both Offert and the MiSTer developers highlight as part of FPGA-based archaeology informed the development of Near's cycle-accurate software-based Super Nintendo/Famicom emulator, bsnes.[37]

In fact, Near argued, FPGAs and software emulation are indistinguishable in capability, differentiable only by the physical features of their respective hardware: "Whether coding in Verilog for FPGAs, or C++ for PCs, these languages are both Turing-complete. That is to say, there is absolutely nothing that one can do that the other cannot. It's only a matter of computational resources and energy cost."[38] In other words, given enough computing power and attention to detail, there is no "semantic surplus" to discrete devices that cannot be recaptured in software.[39] They even outline a case where FPGA hardware limitations *prevented* perfectly accurate recreation of the SNES. It is therefore these material aspects—speed and related concepts like input latency as well as pricing of both chips and power usage—Near concludes, that can distinguish FPGAs from software-based emulation.

These are, of course, the same concerns dominating evaluations of modern CPUs. As such, they fall well within the ambit of Moore's law as both a technological and economic phenomenon. Cheaper, more specialized hardware makes FPGAs an attractive option in comparison to PC emulation. Given the immense performance cost of accurate reproduction, circuitry that can be reconfigured to behave more like original system hardware can offer several advantages over software designed to run on modern commodity CPUs. This is a particularly salient aspect of FPGAs if we are skeptical about general-purpose processors maintaining their past rate of improvement. While from a performance perspective debates over the emulation of older systems are largely scholastic at this point, the issues they raise are of great interest with respect to more recent consoles.

It does not make sense, for example, to purchase a top-of-the-line gaming rig just to emulate contemporary hardware. While sometimes possible, the former is liable to be an order of magnitude higher in price than the latter and, further, potentially introduces the kinds of inaccuracies that cycle-accurate obsessives take pains to avoid. Yet what happens when such systems themselves become as obsolete as the Super Nintendo, if computing power in general has not significantly surpassed them? Beyond theoretical questions about the legitimacy of preserving a piece of software by emulating it on non-"original" hardware, 100% utilization raises practical problems about how to do so with acceptable levels of accuracy.

Perhaps this is one reason, for all of the console industry's historical opposition to the PC model, commentators have begun to speculate that this generation of custom gaming hardware may be the last. The dilemma the business will then face resembles that found in many other areas of a thoroughly digitized media industry ossifying under late-stage capitalism. It can either move into the nearly unpreservable realm of cloud-based services, where corporations provide not individually executable copies of media but only access to centrally hosted instances that can be revoked at any time, or make a comparatively more democratic acquiescence to the longstanding model of personally run software on a standardized architecture, with all of the exposure to piracy and loss of control that has proven to entail.[40] Barbarism or socialism, then, if you will.

It may be argued that full access to the source code for these applications would be the most democratizing step. This evokes the long-standing—though often from a practical perspective erroneous—association between the open source movement and communism. Yet few groups have proven more adept at working *without* the source—that is, at manipulating ostensibly incomprehensible machine code—than the gaming and emulation communities. In these subcultures, programmers regularly create translations and enhancements to "closed-source" games via binary patches that modify executables directly.[41] If it is true that 100% utilization will revitalize the need for coders to operate at the lowest levels of the stack, these projects and others like them will have proven important precursors. Retrocomputing, in this respect, may ultimately end up being much more than a nostalgic exercise. It may foretell the return of a more streamlined model of development for both software and hardware, one attuned to both the particulars of hardware and the optimization of software to maximally exploit it. As Pesce suggests, this would shift emphasis back to human developers, whose skillset would need to extend to these lower computational layers, at which it is considerably more difficult to program.[42]

ASICs: Computation at the Point of Collapse

Such operational synergy already exists for a limited number of highly desirable use cases. In them, we come much closer to the fundamental boundaries of silicon-based, discrete electronic computation. However much of a performance improvement FPGAs may offer over standard CPUs, the lower level of hardware configuration their software interfaces provide do not represent a maximum of efficiency. Indeed, this is due at least in part, as we have previously seen Michael Conrad argue, to their very programmability. Application-specific integrated circuits, by contrast, deliberately sacrifice this mutability for even stronger performance gain. The 2011 National Research Council report, for example, postulates that, at the close of the twentieth century, ASICs consumed a thousand times less power than commodity silicon, while a 2007 effort at evaluating their performance against FPGAs showed, at minimum, a threefold speed advantage.[43] As their name suggests, ASICs are chips designed for a particular use case, what in Moore's (and Engelbart's) original analysis was called a "function."[44]

The rise of the ASIC is intimately tied to both Moore's law and the development of the CPU, including the Intel 4004 microprocessor that forms the jumping off point for "There Is No Software." The roots of what we now think of as general-purpose CPUs began in ASICs, when "Moore's law meant that the existing [mid-1960s] product line of a few simple, high-volume standard ICs would be replaced by many specialized, low-volume, medium and high-complexity ICs. Recognizing that this would mean a fundamental change in Fairchild's way of doing business, Moore assembled a . . . group of CAD specialists."[45] This differentiation eventually recongealed around what became (and likely still are, the centrifugal effects of approaching 100% utilization notwithstanding) the major computer chip categories: CPUs and RAM.[46] For example, of the five chips Kay mentioned in the HP-35 calculator, three were memory while the other two split the functions of calculation and scheduling that CPUs combine.[47] Though their history stretches back almost to Moore's original formulation, contemporary ASICs cannot help but carry with them the connotations of the software deployment of this same terminology. There is, in modern-day ASICs, a resynthesis of the originally connected but subsequently divergent software definitions of "application" and "function" with those in the electrical engineering sense.[48]

One of the most prominent, and controversial, uses for these chips today is in cryptocurrency mining. ASICs run the algorithms necessary for producing valid hashes of digital transactions at rates faster than other designs like CPUs, GPUs (which turn out, rather by accident due to their parallelism, to perform this task well), and, ultimately, FPGAs.[49] Hashes are numbers that can be reliably computed from a given set of input data, but which are extremely computationally difficult, if not impossible, to turn "back into" the data they represent. They are often used to verify informational integrity: If your hash of a downloaded file matches one you receive separately from its creator, you can be assured you have the "same" data. Cryptocurrency miners thus test inputs across the problem space to see if they produce specific numerical outputs. Only a fraction of possible hashes will do so. ASICs earn their owners a larger chance of receiving a share of the currency by computing one of these so-called golden hashes faster than competing hardware.

As a thus far ever-shrinking barometer of socially necessary hash-time, both the elimination of abstraction layers and raw hardware improvements are of the essence. In the words of one of the most popular vendors of Bitcoin mining hardware, in contradistinction to GPUs, "ASIC miners are purpose-built to do one thing and one thing it [*sic*] does best, mine cryptocurrencies based on the algorithm it [*sic*] utilizes."[50] Most vendors do not provide even this much of an explanation, simply listing their products' hash rate, currently in the terahash (trillions of attempts) per second range, and energy consumption, measured in joules per terahash. Anyone in the market for such a device, which is typically installed in a data center to accommodate its power, cooling, and uptime requirements, already knows why they want it.

Critical assessments of cryptocurrencies, and particularly Bitcoin, where ASICs are now practically required to successfully mine, frequently emphasize the amount of electricity dedicated to keeping them running, describing such distributed digital assets, either implicitly or explicitly, in terms of waste or *in*efficiency.[51] This characterization has become increasingly accepted, with New York State enacting a law to restrict the use of energy-intensive currencies like Bitcoin and rivals switching to alternative legitimation schemes.[52] In one of the most astute analyses in this vein, Jeffrey West Kirkwood has argued that

> in place of efficiency, [Bitcoin] relies on increased work; in place of simplification, it substitutes complexity. This is a situation . . . that stands to invert prevailing notions of value in the digital economy—and because this value stems from processes that destroy information through irreversible modes of encryption, it destroys traditional notions of meaning as well.[53]

For Kirkwood, Bitcoin is inefficient because of the way that it relies upon semi-arbitrary values, known in cryptographic terminology as numbers used only once, or "nonces." These guarantee the integrity of the currency's ledger—the blockchain—beyond the commonplace use of hashing functions to attest that information is what it is supposed to be. They do so by demonstrating a sufficient amount of computation has been performed to justify a reward for verifying each new block of transactions. This "proof-of-work" encourages competition for this remuneration and in this way makes the monopolization of control over, and therefore the potential forgery of, these records unlikely. This in turn makes the system a reliable indicator of balances and conductor of transactions.

The point of the nonce here is to introduce a fungible data element into the cryptocurrency algorithm. Miners can then manipulate it to produce a golden hash. Nonces are combined with several more straightforwardly determinate data points, including the time of computation, the hash of the most recently verified block, a multilevel hash of all transactions the block contains (known as a Merkle root),[54] and the level of difficulty at which the requirement for proof-of-work has been set (according to the total amount of computation devoted toward mining the currency). This agglomeration is then itself cryptographically hashed until it generates a golden hash, which is then used as part of the dataset to generate the hash of the following block, and so on (thus the name blockchain). As the difficulty increases, the likelihood of a nonce being found—a "golden ticket"—which produces a golden hash decreases, and this fundamental relation maintains the blockchain as a whole against manipulation.[55] As the pseudonymous white paper introducing Bitcoin (written prior to the transformations in hardware concerning us here) states, "The system is secure as long as honest nodes collectively control more CPU power than any cooperating group of attacker nodes."[56]

Since the security of the system is dependent on the commitment of processing power, and the more committed, the more difficult any individual verification becomes, the system itself, for Kirkwood, is literally built on inefficiency. Moreover, he argues, this represents a marked change from industrial and nineteenth-century understandings based in thermodynamics, in which "the exploitation of labor essential to capital accumulation relied, at least in theory, on ever-greater efficiency—a management of entropic mechanical systems and the heat that made work quantifiable from Sadi Carnot's 1824 *Reflections on the Motive Power* onward."[57] In contradistinction, he argues, Bitcoin is emblematic of a larger drive within *post*industrial capitalism to generate value from increasingly complex and wasteful tasks, particularly the kind of data collection and analysis regularly purported to be the source of profit for service-based tech platforms like Google and Facebook. Such systems' wastefulness, he ultimately concludes, contributes to "the only truly nineteenth-century industrial element of its construction: heat death,"[58] or entropic decay.

This analysis resonates with conceptualizations of the relationship between work and labor, a distinction Kirkwood appears to recognize, albeit without elaboration,[59] that I will turn to at the end of this book. Here, however, it is necessary to take up with the more technical aspects of this critique and others, such as Zane Talley Griffin Cooper's, which acknowledges and broadly comports with Kirkwood's but goes further, contending that "proof-of-work systems see information as an outcome of entropy, or expended, irreversible energy use. . . . The waste is the point, the meaning in all of this. In proof-of-work systems, expended energy is given meaning and value, thereby pointing directly to a panacea of energetic relations."[60] This latter sentence, however, seems to me equally applicable to industrial commodity production as it is to cryptocurrency mining. Foreshadowing some conclusions, then, let us first acknowledge that an argument for efficiency does not necessarily translate into one for desirability. As the punk band Dead Kennedys satirically posited in a Swiftian reductio ad absurdum of the neoliberal postwelfare state, atomic radiation can be a wonderfully effective tool for the elimination of the unemployed.[61] At the same time, Jonathan Sterne's archaeology of the MP3 has shown how the format's more efficient use of storage encouraged its adoption over bulkier, uncompressed compact disc audio, despite its lower fidelity.[62]

Which political valences and operational trade-offs, exactly, characterize cryptocurrencies like Bitcoin remain up for debate. While the dominant tendency is to view them as expressions of a right-wing libertarian, anarcho-capitalist ethos—an attitude certainly supported by the number of scams and collapses in the thus far largely unregulated (and to some extent deliberately impervious to regulation) area—there have also been stirrings of a forcefully left-wing defense of the concept.[63] The need for a method of tracking value following the postsocialist transition to a state where communism and (left-)anarchism's goals coincide seeming as plausible to me as the possibility that Bitcoin itself will not, in fact, be this method, I will constrain myself here to a different line of inquiry. It is the question of what *efficiency means* and, in particular, the meanings ascribable to computational efficiency inseparable from a consideration of 100% utilization. This will necessarily touch on questions of ecology, for the relationship between waste and efficiency is a complex one. Our increasingly efficient computational hardware, paradoxically, has given rise to greater environmental concerns related to its use.

To address this question, the first of two conceptual pairings requiring further disambiguation is the relationship not between work and labor but between work and *heat*. As Adam Greenfield notes in his discussion of Bitcoin, "Computation . . . isn't some species of magic: every last bit that's flipped in a server somewhere is work, in the strict physical sense. Like all other forms of work, it requires the application of energy, and generates waste in the form of heat."[64] The existence of heat in a process is thus certainly proof of inefficiency, as Kirkwood notes. This does not, however, indicate, as he suggests that it did for industrial-era factory labor, that "greater heat meant greater inefficiency."[65] Cooper's contention that heat, rather than work, is the salient factor in Bitcoin mining makes a similar conflation. In the case of an engine lifting a weight, a classical thermodynamic example we will have cause to return to, two identical engines lifting two identical weights simultaneously will raise the temperature in a room more than one engine lifting both weights one after the other. Yet, thermodynamically, these two events are equivalent in terms of energy input, work done, and entropy lost. It is for this reason, not due to greater inefficiency, that data centers must be cooled much more aggressively than personal systems. More computers are doing simultaneous work within a densely packed amount of space. In this way, racks of servers recapitulate at a macrophysical scale the material problems encountered by semiconductors at the microphysical. However, if there has been a break between an industrial and postindustrial mode of production, it has been not in the abandonment of efficiency but its opposite: the complication of any linear relationship between heat loss and work done. This is a paradoxical complexity precipitated in no small part by Moore's law.

We can express this contradiction in the following way: ASIC miners are some of the strongest computational drivers of both energy consumption *and* energy efficiency. This is because energy consumption actually goes *down* as transistors get smaller and computing power—work—correspondingly increases. As famed electrical engineer Carver Mead discovered in the late 1960s, prompted by a query put to him by Moore about the ultimate physical limits of transistors, "contrary to all the current lore in the field, we could scale down the technology such that *everything got better*. The circuits got more complex, they ran faster, and they took less power."[66] This was already largely noted in Moore's foundational article, where he dismissed concerns about heat: "Shrinking dimensions on an integrated structure makes it possible to operate the structure at higher speed for the same power per unit area."[67] The case of concentrated computing power, in ASICs or data centers, and especially ASICs *in* data centers, thus represents a particularly significant angle from which to approach this dynamic. As Cooper contends, there is both little difference between mining and other high-intensity computational processes (they are all effective procedures, in Turing's sense) and a very real way in which the adoption of ASIC miners has contributed to the overall data center efficiency these other applications now enjoy.[68] At the same time, his focus on heat over work is misleading in the sense that it does not fully address the necessary but in some sense ineffective work performed by proof-of-work systems, or indeed any computationally intensive process that is not 100% deterministic. It is not really the case that "the exhaust is the *proof* of the *work*,"[69] or at least not the bulk of it, particularly as efficiency gains, as long as they last, continue to drive this exhaust down in proportion to the work done. Strictly speaking, more energy is "wasted" in completed hashes that do *not* result in golden tickets than in the entropy lost to both these and winning hashes together.

For Anne Pasek, Moore's law, coupled with recent exponential advances in cooling and idling servers even as it has started to falter,

> sets ICT [Information and Communications Technology] apart from other industries; it has to some significant degree decoupled its rate of economic growth from its most obvious forms of environmental impact. As such . . . it suggests that climate and capitalism might not be fundamentally at odds and that a turn to a more digital economic base may be the way through the environmental challenges of the coming decades.[70]

However, as with corollaries of Moore's law such as Wirth's "law," specifying the tendency of software complexity to strain hardware capabilities, this is an example, as filmmaker and media scholar Laura Marks describes, of a phenomenon of keen interest to environmentalists concerned with the carbon impact of digital technologies like (as she is) streaming media, "the Jevons paradox: more efficient technologies often encourage greater use of a resource, reducing or eliminating savings."[71] Pasek's analysis, while not invoking it directly, concurs, noting that "while the carbon and energy intensities of data storage and network transmission continue to fall, the total stock of the industry's infrastructure and energy needs is still creeping upward," ultimately prompting her skepticism of capitalism's ability to stave off this trend.[72] What might seem at first to be one contradiction—computation costs energy and thus an increase in one invariably leads to an increase in both the other and waste heat—is revealed instead as another—the Jevons paradox—thanks to a remarkable property of Moore's law.

In this sense, both 100% utilization and, as Pasek describes, a separate but related minimum to how *little* energy a powered-on, but unutilized, system can use may present a capitalist technology industry with difficult choices between a dramatic increase in the environmental impact of data centers around the globe and a rationing of the computing power they contain.[73] Yet, as this example also shows, an increase in heat does not necessarily represent a decrease in efficiency. We must be mindful of the amount of work we are getting in return. If we produce, say, twice as much heat but gain four times as much computing power, we are ahead in the efficiency sweepstakes, and our results are measured by the latter statistic more than the former. When systems like cryptocurrency miners are running computations constantly—100% utilization in the sense of a specific processor or set of processors but not necessarily 100% utilization in the broader theoretical sense—efficiency gains scale with power usage in what Cooper calls a "borderline-indexical" relationship.[74] Since these processors are never idle, they are never wasting power waiting on direction.

This is the definition of thermodynamic efficiency, at least when considered in terms of energy: an increase in energy available to a physical system for work—or, put negatively, a decrease in the amount sacrificed to entropy—per unit input. As such, far from representing a supersession of nineteenth-century physics, the metric provided by ASIC miners, joules per terahash, is perfectly thermodynamic. It allows prospective buyers to understand not only how much energy is required to power these systems but also how much is being lost. The smaller this number, the more efficient the system. Although in practice usually much higher, the second law of thermodynamics establishes a maximum efficiency of zero increase in entropy. Since a decrease in entropy—an output of work beyond the amount of energy fueling a system—would amount to creating energy ex nihilo and thus violate the first law of thermodynamics, systems can only produce as much work as they receive in energy, or less. The former case is the best case, also known as a reversible process, and generally impossible under real-world circumstances, representing a theoretical boundary.[75] Otherwise, "in all irreversible processes," as noted theoretical physicist Max

Planck described, we encounter the familiar definition of entropy: "A certain amount of work is lost, viz. the difference between the maximum work to be gained (the decrease of the free energy) and the work actually gained."[76] All efficiency is geared towards minimizing this quantity, and thus *decreasing* the ratio between energy used and work done—that is, decreasing the amount of heat expended per unit of work.

Engineered toward a specific task, ASICs can optimize their *efficiency*, their ratio of joules in to output produced, even further beyond what is to be gained by decreasing semiconductor size. This makes them especially attractive at a time when the future of Moore's law is in doubt. It also makes them another avenue for improving data center efficiency. Rather than focusing on general computation, servers could become more specialized to individual tasks, and they are.[77] Moreover, ASICs offer opportunities to push computing *performance* in specific areas past what other silicon can offer, as in cryptocurrency mining. There, it is equally the case that technological developments in efficiency have fed the sector's Stakhanovist drive to turn computing power into money as it is that competition within the sector has driven usage. ASICs ride the cutting edge of Moore's law precisely because of the advantage they offer over more flexible hardware. The overall effect has thus been one of increasing numbers of hashes and decreasing energy cost per hash—resulting, consequently, in less heat lost per unit of useful work.

If one aspect of a critical assessment of ASICs rests on the distinction between thermodynamic work and heat, and particularly an understanding of their distorted interrelationship due to Moore's law, the second binary to be considered must be that between encoding and meaning. For Kirkwood, Bitcoin "relies on destroying meaning. It employs cryptographic tasks of escalating difficulty that are not intended to secure and protect meaning—as is the case in the use of encryption for clandestine services or internet security—but to destroy it in processes of computational excess whose reward is an accursed token of cryptocurrency."[78] Since nonces force miners to search for an arbitrary value to produce a specially formed hash, they raise the computational work of verification far beyond that of the single hash necessary to validate the rest of the data fields.[79] However, two assumptions in this statement bear further consideration. The first of these is that cryptography, more broadly writ, has something to do with protecting *meaning*, which could come as a surprise to at least some poststructuralist attempts to reckon with the practice.[80] The second is one that Cooper points out when he writes that "perhaps we should ban proof-of-work, but in doing so, we must understand that . . . it merely asserts a value proposition, making a cut that separates good data from bad data. This cut inevitably blames proof-of-work for the broader toxicity of data's energetic relationships, which proof-of-work tends to bring into relief."[81] It, is in other words, the assignation of not value but a value *judgment*—indeed, meaning—to the result of successfully hashing the blockchain.

The issue with the first supposition is that *all* informational processes and formulations take place, regardless of medium and following Claude Shannon's cardinal intervention, irrespective of meaning. As a study of signals or messages, information theory dictates that "frequently the messages have *meaning*; that is they refer to or are correlated according to some system with certain physical or conceptual entities. These semantic aspects of communication are irrelevant to the engineering problem. The significant aspect is that the actual message is one *selected from a set* of possible messages," the breadth of this set being defined as its entropy.[82] The consequences of this statement are vast, incidentally rendering perplexing Cooper's emphasis on "the semantic rendering of information and entropy" as a site of rupture between materialist and symbolic understandings of these concepts.[83] As I have explored elsewhere, in introducing the concept of a (binary) sign system denuded of signification upon which nearly all of our contemporary communications are predicated, information theory shifts the ground for political engagement from language to action.[84] A theory of 100% utilization logically follows from this, but in doing so it also illuminates some further consequences of this development.

If meaning is a priori absent from an informational system, then there is nothing about hashing algorithms or ASICs that could further contribute to its eradication. It is true that hashes, by themselves, almost definitionally lose information in comparison to their inputs, but this does not make them computationally unique in any way. "Lossy" compression, for example, of the type applied to audio files is another example of this. Certain aspects of "uncompressed" digital sound deemed imperceptible may not be considered significant to these algorithms, and thus two source files identical save for these aspects could produce identical compressions from which this "superfluous" information would be unrecoverable.[85] More directly relevant to our concerns, Kittler himself argued that compilation is akin to a form of hashing.[86] This is obviously a much more destructive form of loss since, unlike verifying downloads or blockchain blocks, this method can be deliberately employed to obscure, rather than support, more human-workable renderings of how software operates. Yet compilation, by turning plain text into machine instruction, is critical to making most software *do* anything at all! As Chun has suggested, this elimination of information is more of a Heideggerian challenge to the metaphysics of presence than straightforwardly wasteful: "*Source code only becomes a source after the fact* . . . only through its destruction, through its simultaneous nonpresence and presence."[87] It is not, then, a defining feature of postindustrial capitalism that value is created via "processes whose purpose is to make things *less*, not more, readable and therefore *less*, not more, efficient."[88] At these much more fundamental levels of computation, well in advance of the kinds of processing-intensive algorithmic value extraction that, for Kirkwood and Cooper, is exemplified in Bitcoin yet also visible in other high-performance computational processes, destructive illegibility and efficiency are not at odds but rather complementary.

Assessments of the destruction of information in computationally optimized processes like Bitcoin mining that run counter to those of compilation, then—evaluations of informational loss as wasteful or harmful instead of productive—rely, as Cooper suggests, on a *judgment* of value. Moreover, it is one made in contradistinction to the value attributed to Bitcoin as a financial instrument.[89] As such, they are meaningful, but not informational. Speaking informationally rather than thermodynamically, hashing *reduces* entropy: A Bitcoin block header is 640 bits, but its hash is only 256.[90] At the same time, however, a golden hash is—as its nickname suggests—extraordinarily "valuable" in informational terms. Its rarity in fact regularly exceeds that of the 2^{32} possible nonces comprising the problem space. If none produce the desired type of hash, one popular solution is to re-search the space at a different time, as little as one second apart, effectively rendering a block's timestamp part of the nonce.

A golden hash is thus an exceptionally rare informational event and the ultimate proof of work. As one of no more than a handful of syntactically valid responses, it has low relative entropy within the highly entropic overall hash space. Relative entropy is "the ratio of the entropy of a source to the maximum value it could have while still restricted to the same symbols," and, as representative of "the maximum compression possible when we encode into the same alphabet," is in fact tied to the exchange of calculative power for smaller and/or more secure messages.[91] This is because, as alluded to in the previous chapter, "in general, ideal or nearly ideal encoding requires a long delay in the transmitter and receiver" for "reasonably good matching of probabilities to corresponding lengths of sequences," a rare concession to computing time and therefore materialism from within the symbolic realm of information theory.[92] Symbolic order here is the product of thermodynamic work and thus the dialectical counterpart of not just informational but physical entropy.[93]

This relation holds even as efficiency drives down the ratio of heat to computation in a process that approaches—but never reaches—sublation. Regardless of how unlike its inputs it may be, a golden hash is the infrequent and highly structured product of a physical process. It is this, rather than its symbolic qualities, that fundamentally gives it value. It is the product, in other words, of *work*, if not necessarily (much) labor.[94] A golden ticket will thus have a value associated with it both derived, like the physical commodity it virtualizes, from the components that go into producing it—here circulating capital (energy), fixed capital (hardware), and some modicum of variable capital (labor, e.g., in configuring and maintaining the system)—and varying, also like material commodities but much more so like financial instruments, according to the perception of this value, a perception predicated in no small part on the fantasy that value can be created by machines rather than human labor.[95] To call these tokens "accursed," then, or their power draw "dangerously wasteful" and "monstrous,"[96] is to attempt to influence this latter aspect only, and to do so on fundamentally moral, rather than operational, terms.

Following Cooper, who as we have just seen both acknowledges this and contributes to it, a value judgment on Bitcoin is an attempt to establish the *meaning* of the technology. It is only on the basis of such a judgment that we can declare cryptocurrency mining uniquely wasteful, in contradistinction to any other work, computational or macrophysical. Such a judgment has nothing to do with the thermodynamic definition of "useful work," however, which, from Carnot's foundational characterization on, has often luxuriated in this contradiction. Carnot offered "the expression motive power to express the useful effect that a motor is capable of producing. This effect can always be likened to the elevation of a weight to a certain height."[97] It is not necessary that such a raising *be* useful, only that it *could*.

The NRC report describes computation similarly: "Computer performance has historically been defined by how fast a computer system can execute a single-threaded program to perform useful work."[98] As Pasek, Cooper, and their coauthors of *Digital Energetics* note, "Physics . . . argues, simply put, that work is the transference of energy. This definition allows muscles and combustion engines to be jointly apprehended in joules of force expended, effecting a kind of posthuman commensuration between organic and inorganic forces. Signal processing and information theory, formally agnostic to the specific materializations of a given symbol, are inheritors of this tradition."[99] I shall have occasion to explore at length this conflation of the human and inhuman alongside the distinction between labor and work toward the end of this book, but here it is worth remarking that is not simply the *medium* of work but its *purpose* that is unimportant to physics.[100] In the eyes of thermodynamics, driving to the grocery store and doing donuts in the parking lot are equivalent activities: Energy is consumed, work is produced, and heat is lost either way. In this sense it is also, like the information theory derived from it, agnostic to meaning, just as Bitcoin itself is "agnostic to energy consumption."[101]

As such, from the perspective of 100% utilization, the proper ground for understanding ASICs is not their production or destruction of meaning or how we feel about the way they are being used. It is, rather, their operation as a thermodynamic process, a process they share with every other computational procedure and much more besides, and its imbrication with the profit motive. A lot has been made of not just the supposed inefficiency but the power-hungry nature of proof-of-work systems. Most famously, a 2014 paper by Karl J. O'Dwyer and David Malone estimated Bitcoin energy usage at the time rivaled that of the country of Ireland.[102] Kirkwood, citing the Cambridge Bitcoin Electricity Consumption Index, notes a more recent claim vis-à-vis Belgium, while Cooper draws upon the Bitcoin Energy Consumption Index to offer up Ukraine.[103] As the former scholars write, "the major limiting factors in Bitcoin mining are the hash rate of hardware and the cost of running this hardware. . . . There is a trade off between two time varying factors: first, the energy cost of discovering a block. . . . Second is the cash reward for discovering the block."[104] Users and developers of these systems are highly motivated to keep electricity usage down; energy consumption and related thermal management costs eat directly into their profits.[105] In this way, "the competition created in mining for Bitcoin has lead [*sic*] to a situation where in order to be financially viable the hardware has had to become faster and more energy efficient."[106]

The ongoing conception of cryptocurrency mining in terms of both computational and economic efficiency, then, has led not to the supersession but rather the *perpetuation* of supposedly outdated nineteenth-century precepts. This has happened, further, despite the complication of the relative relationship between heat and inefficiency brought about by Moore's law. However, this not a question of "assuming a limitless extension of older industrial modes of exploitation"[107] but one of interrogating their very real limits. A bigger question than how much energy these systems are using, or how much heat they are generating, might therefore be what happens if and when material improvements to electronic components no longer develop. This situation will eventually arise no matter our sentiments on their use. The end of Moore's law, as we have already seen, threatens the feasibility of data centers, but there the situation is complicated by the presence of both busy and unoccupied servers. In cryptocurrency ASICs we have an example of computational optimization pushed to its limit, and the answer to what happens when they run out of growth could be less obvious than it may seem.[108] Conditions of 100% utilization may change not only *how much* energy and computation is used, but *how* it is used at all.

This is a question separate from, but related to, that of the most identified internal limit to the Bitcoin system: What happens when the total allowable number of bitcoins has been distributed to miners? As O'Dwyer and Malone note, "After such time it is imagined that the network of miners will continue mining but will do so in order to gain processing fees," also distributed in Bitcoin, but "the current standard reward [for mining] . . . is considerably larger than the current or historical average transaction fees."[109] Miners are already in a sense chasing a law of diminishing returns, as the "block reward" of coins for a successfully accepted hash halves, in a rough inverse of a slackening Moore's law, approximately every four years as the number of available bitcoins to distribute dwindles.[110] This exponential depletion has been counterbalanced by the currency's dollar valuation, which remains substantial despite its variability. Once the remaining block rewards have been distributed, however, even if the comparatively paltry transaction fees were raised substantially, this would not begin to make up for sunk hardware costs. Already, the price of both ASICs and the amateur's mining technology of choice, GPUs, drop when the token's value declines, an expected result if the laws of physics remain a determinate factor in the enterprise.[111]

Now consider a situation in which Moore's law ends—that is, one where eventually both hardware speed and energy efficiency are fixed. Under conditions of 100% utilization, mining capability would become a direct function of the amount, rather than the quality, of hardware thrown at it. Consequently, output would also be a direct function of hardware quantity times energy cost, significantly complicating what Kirkwood identifies as the way that computing-heavy processes like Bitcoin mining "intensify inequalities based on access to hardware."[112] Although it appears we have not yet reached this point, the 2011 NRC report already anticipated that "the growth in the performance of computing systems—even if they are multiple-processor parallel systems—will become limited by power consumption within a decade."[113] If and when this happens, as with data centers more broadly, the Jevons paradox will be replaced by the much more damaging linear relationship between energy and output it staves off. Paradoxically, however, since the Bitcoin algorithm adjusts its difficulty based on the total amount of computing power thrown at it, the difficulty of mining could *diminish* as people stopped throwing their computing power at what would amount to an utterly, rather than gameably, stochastic lottery.[114] Such a development would result not only in a decrease in the price of Bitcoin but also its increasing instability as a peer-to-peer mediator of transactions. A shrinking pool of miners would make it easier for a malicious actor to dominate its ledger, further depressing if not destroying the token's valuation.[115]

It is hard to imagine the diverse array of computing power leveled at cryptocurrency mining persisting considering either one—let alone both—of these approaching limits, logical or material. Neither of these, however, have anything to do with any subjective assessment of the value in keeping the technology running. Real solutions to the problem of increasing energy usage, a common refrain holds, require methods of verifying transactions less computationally intensive than proof-of-work: "Unless bitcoin changes to proof-of-stake the bubble is likely to end either in a financial or environmental catastrophe."[116] Yet proof-of-stake verifications, which are largely unproven at scale, have their own issues, most notably the danger they could effect a recentralization of (in)famously distributed currencies.[117]

For Bitcoin specifically, which remains committed to proof-of-work and thus dependent on ASICs, if such a transition was ever made it would likely be for internal, rather than external reasons. That is, a switch to proof-of-stake would be made not to save energy directly, but to save the system from rising power overhead alongside stagnant efficiency.[118] Just as rising interest rates constrain the value of highly speculative assets like cryptocurrencies, when computing power ceases to improve, mining Bitcoin will no longer be economically workable no matter the environmental externalities. The same could even turn out to be true for cloud providers, already representative of a significant and disturbing consolidation in comparison to the Internet of decades past.[119] Under conditions of 100% utilization, competition may dry up as the cost of adding new users increases linearly rather than logarithmically.

Efficient Causes

The existence of custom silicon and the thinning razor's edge upon which it skates is, as with FPGAs, a question of performance rather than logic or judgment. Scholars like Chun may take care to note that enshrining logic in hardware rather than software does nothing to obviate the symbolic—and thus, ultimately, deconstructibly incoherent—nature of these procedures as *logos*.[120] Miners and others who choose ASICs over the adaptability of software, however, are driven by material outcomes rather than logical consistency, *even when measuring performance vis-à-vis this or that logical system*. While the algorithms behind its mining are robust enough that the Bitcoin system has not (yet?) been broken, one need not subscribe to the countervailing notion that mathematics offers some sort of ontology to note that efficiency, and not logical coherence, is what is of essence here. This drive to turn more programmable but less optimized software into high-performance, energy-efficient hardware in this and any other arena may well prolong gains. It will not, however, resolve the internal contradictions driving economic reliance on computing over labor power.

Dialectical resolutions to what is ultimately the contradiction between consumption and production—or, indeed, exchange value and use value—need not only take the form of a conversion to a new physical foundation, however, as engineers and proponents of Moore's law hope. They can also be driven by a change in the balance between technological and social forces. This is the stage where meaning, expunged by seventy-five years of informational and computational efficiency, can and should properly enter. We could decide, for instance, to forgo the massive amount of pointless, CPU-hungry advertising with which the online publishing industry lards its web pages—the true source of Facebook and Google's wealth, rather than the data mining that feeds off of it—or the enormous state surveillance apparatuses collecting, storing, and decrypting the world's internet traffic in gigantic data centers that nobody has voted for in either the ballot box or marketplace.[121] We could, instead, direct the savings in money, energy, and computation toward more optimal (indeed, efficient) designs for the tasks we *do* place significance upon accomplishing electronically—potentially including some type of decentralized currency.

An assessment of cryptocurrencies at this level could properly account for the meaning of not only the waste our systems produce, as Cooper argues, but the work they conduct. As one group of computing researchers has argued, "We have to put bitcoin's energy consumption into context: how much it consumes compared to other uses of computing power, what value or benefit mining creates (and for whom), and consequently what the appropriate alternatives might be."[122] Similarly, the NRC report argues that "the net environmental effect of more (or better) computing capabilities goes beyond simply accounting for the resources that server-class computers consume. It must also include the energy and emission savings that are enabled by additional computing capacity."[123] A good-faith evaluation of proof-of-work-based currencies would have to account for not just their relationship to one country's energy usage, but the total consumption of "legitimate" finance operations *globally*.[124] This would include everything from the computationally intensive server farms associated with financial trading to the less active consumption, but much more numerous existence, of always-on ATMs, Bloomberg terminals, point-of-sale processors, and bank employee desktops spread around the world, not to mention the cooling costs (including office cooling) associated with these, and more.[125]

Even more fundamentally, comparing the energy consumption of something like the explicitly antistatist Bitcoin with that of relatively small countries rhetorically contrasts the two. Doing so not only delegitimizes the former, it validates the latter, and this despite the long-standing critical conception of nations as in some appreciable sense phantasmatic.[126] "What is the burgling of a bank to the founding of a bank?"[127] And what is the waste of Bitcoin compared to the profits financial institutions derive every second from the money we deposit, yet they accrue to themselves? Moreover, what juxtaposition of stateless cryptocurrencies with the transcontinental patchwork of nation-based economies would be complete without also accounting for Belgium's brutal historical role in the exploitation of the Congo—a legacy that persists as its natural resources are now critical to the production of capacitors, electronic components used, ironically, to store energy—Ireland's neoliberal transformation into a tax haven for global technology companies, or Ukraine's disastrous reduction to a pawn in twenty-first-century sphere-of-influence squabbles?[128]

A dialectical consideration of these conflicting financial systems, liberal-democratic and anarcho-distributed, is necessary to avoid any misprision that one or the other is either entirely harmful or wholly advantageous. Rather, both are organized around substantial vulnerabilities, while each offers protections the other does not (and often explicitly rejects). It is tempting to contrast the two by recourse to their structuring commitment, the operational necessity of exploiting either labor or computation. Yet, in taking up such an evaluation, we would also move closer to articulating a synthesis, a better means of distributing value *in general*, particularly for an age of slowing computational dynamism and rising labor power. That is, we could take the efficacies of these existing processes as a benefit of the capitalist mode of production and rebuild them toward ameliorating, rather than perpetuating, its vast liabilities.

Such urgently needed changes would necessarily entail a collective project largely free of the politics of individual responsibility that so often characterizes the conjunctural compromise between environmentalism and the systems of governance tasked with perpetuating a political-economic regime based upon obsolete presumptions of financial and technological growth alike. As Sean Cubitt argues, "This individualist approach shifts the burden of toxicity away from industry toward consumers, in a move that parallels the shifting of debt and austerity from failing economic and political elites to the already overburdened poor."[129] At the same time, however, such a realignment would resist "actual degrowth"[130] in favor of a more purposeful allocation of the already abundant resources at our disposal. It would also encourage the pursuit of efficiency as a collective, rather than privatized, goal. Rather than becoming increasingly "conscious" of our utilization of electronic resources, streaming or otherwise—the NSA would love it if you used less intensive cryptography!—such a movement would reach far beyond questions of consumption or optimization to upend the cybernetic order substituting silicon for human potential. In doing so, our labor would become meaningful in ways it has never been before, while our machines would continue to carry out their tasks—albeit toward very different goals.

If thermodynamic efficiency is impartial toward the ends to which its systems' increasingly impressive marshaling of energy are put, we need not say the same for those who first posited it. Carnot, in fact, concluded his treatise with just such an acknowledgment of their significance:

> We should not expect ever to utilize in practice all the motive power of combustibles. The attempts made to attain this result would be far more hurtful than useful if they caused other important considerations to be neglected. The economy of the combustible is only one of the conditions to be fulfilled in heat-engines. In many cases it is only secondary. It should often give precedence to safety, to strength, to the durability of the engine, to the small space which it must occupy, to small cost of installation, etc. To know how to appreciate in each case, at their true value, the considerations of convenience and economy which may present themselves . . . to balance them properly against each other, in order to attain the best results by the simplest means: such should be the leading characteristics of the man called to direct, to co-ordinate among themselves the labors of his comrades, to make them co-operate towards one useful end, of whatsoever sort it may be.[131]

Noting in passing that the originator of thermodynamics as a field of study was perfectly comfortable with trading efficiency for security, the challenge now confronting us is to combine the judgment hoped for in this passage with the distributed nature that computational networks like Bitcoin have so far only demonstrated in capitalist terms. Rather, in other words, than being governed by liberal technocratic engineers eager to put our labor to work effectively, we must find a way to redirect the outputs our machines already efficiently produce toward our desired labors. These are the things that only we, and not our tools, can do—and decide to do—together. To coordinate among ourselves in comradely accord toward an end worth pursing with all the efficacy we can deliver will represent a true optimization of collective power unattainable under the capitalist dichotomy of use and exchange. In a perverse inversion of standard narratives of technological determinism, it may not be scientific progress but the upsetting of scientific progress, in the form of an end to Moore's law, that brings us to the point where such a reconfiguration becomes possible.

At the same time, bleaker alternatives remain. Indeed, as we approach 100% utilization, any move in the direction of a more positive solution finds itself stymied by an intensification of the disparity between work and labor that cryptocurrencies model. The pursuit of efficiency considered here in terms of hardware, its incorporation of software, and the approaching limitations to its material trajectory has, dialectically and paradoxically, given rise to highly irrational changes in what remains as software and its relation to hardware. Specifically, programming as we have typically understood it is in the process of being rejected in favor of the unreliable, and ultimately impossible, simulation of brain activity. This is a computation-intensive process that may not find as abrupt a terminus in the end of Moore's law as proof-of-work tokens. The need to train neural networks is potentially never-ending, given the unique advantages and applications of each model. Many ecocritics are therefore correspondingly shifting their attention from cryptocurrencies to artificial intelligence. As companies we have already encountered, like Cerebras, indicate, hardware designers are happy to feed the growing appetite for thoughtlike algorithms with power-hungrier silicon. Yet we have not, so far, considered the radical software approach such chips are optimized for. This methodology, which promises to make our machines more sophisticated, is gaining traction at the same time an increasingly precarious workforce finds itself directed toward ever more mechanical tasks. This does not signify that all is lost, however. Instead, such increasingly desperate attempts to cope with an approaching limit to computational growth conceal an impending crisis for the governing classes directing them, and a corresponding opportunity for those whose labors have been forced to accommodate such recklessness.

4 Convolute N(eural): Overused AI and Underemployed Minds

If specialized hardware chases diminishing performance due to tightening material constraints, increasingly appropriating the features and functions of software to do so, software itself changes to meet these same needs. The contemporary push for "artificial intelligence" is one of the most substantial ways it has attempted to do so. AI has rapidly become one of the most buzzworthy computing terms of the 2020s. This development has in turn given rise to new forms of hardware. However, unlike the circuitry examined in the previous chapter, these physical commodities are driven into production more by newly energized computational techniques than a desire to increase the performance of traditional applications. As major tech companies tout specialized chips designed to facilitate machine learning applications, pundits expound upon the ways it will change our lives, musicians make records deploying some version of the technology, and a new generation of scholars from the social sciences and humanities position themselves as gatekeepers to a realm of "ethical" AI, one which promises to rid this new demesne of the prejudices that have bedeviled its all-too-human architects.[1]

As with previous moments in digital technology such as "multimedia," "the network computer," "cloud computing," "intranet," and "dot-com," the term AI is both enabled and hampered by a certain imprecision, its status as a floating signifier that simultaneously means more and less than it indicates. This symbolic ambiguity, deployed in support of a software technique that purportedly does away with symbol manipulation in favor of a simulation of brain activity, has proven quite beneficial for the discourse(s) surrounding the technology. Depending on who you read, AI is a Promethean revolution, the key to tech giants' ability to capitalize on all the data they have collected about us, or a disturbing laundering of persistent social and racial biases into purportedly "objective" algorithmic processes.[2] These technologically minded approaches from critics are mirrored by an increasingly socially oriented bent in many of the field's key technologists. AI experts have become increasingly interested in teasing out its bigger-picture implications, going so far as to ruminate philosophically on the potentially life-threatening problems that await us if and when we—for some reason—pass control of our governance functions over to an all-powerful computer built on the technology.[3]

This range of approaches, while individually offering key insights, tends to share one common assumption: that AI represents a substantial future for computing and, as such, must be reckoned with before its destiny gets out of hand. This premise is accepted, at least implicitly, by even most critical engagements with the field. Many of these crucially contest the political-economic legacy behind the conceptual developments underpinning the technology,[4] a facet I will also have occasion to touch on. But in so doing, they can also take for granted the technology's legitimacy. Others, meanwhile, argue that we must attend to the qualitative increase in computational capacity that AI represents as automation becomes autonomous, moving computation beyond human cognition and into a faster, affective realm where "there are no limits to the epistemological production of knowledge—or understanding and reasoning—that may lead to the evolution of another phase of algorithmic automation."[5] Such conflicting visions of what this future is and how to grapple with it nevertheless frequently disavow the specificities of the technology's physical operation in favor of larger societal narratives about its implications. Such analyses are to some extent enabled by the standard elements of techno-hype that routinely replace material processes with utopian and dystopian conceptions alike. In this case, however, they are also aided by one of the most unique features of modern AI implementations, particularly, the convolutional neural networks (CNNs) upon which they are frequently based. As many primers on the subject will note, it is nearly impossible to for us to really comprehend how they function. In Will Knight's terms, AI systems "have programmed themselves, and they have done it in ways we cannot understand. Even the engineers who build these apps cannot fully explain their behavior."[6]

Why would major tech corporations and scientists embrace such an explicit shift away from the seemingly logical roots of computing? My contention is that the answer to this question can be found in the way that CNN AI, as a software paradigm, is better structured than more conventional programming methods to take advantage of the moves toward parallelism and custom hardware we have considered previously. As the speed of single-core processors has plateaued, AI approaches based on massively simultaneous calculations have become more attractive, despite their rather significant limitations. This development has been hastened by the discovery that special-purpose silicon—particularly the graphics processing unit—has proven unexpectedly effective at the mathematical operations undergirding simulations of neural activity. Here, too, the technical developments necessitated by a looming end to Moore's law can be seen as operating in inverse and paradoxical relation to social trends. While our treatment of computerized "brains" moves away from the ratiocinative straitjacket of formal programming, our view of people's intellectual capacities becomes more instrumentalized. This is true even at the level of higher education, where the acquisition of employable skills increasingly trumps critical inquiry. To comprehend this situation, it is crucial to consider the history of AI in its relation to both software and hardware.

As we have seen, scholars like Wendy Hui Kyong Chun have argued for some time that a lack of perfect understanding is inherent in even the language-based computer programming models that have thus far been the dominant mode of electronic computational production since the invention of the compiler. In this line of thinking, the close correlation between (programming) language and (computerized) action or execution cannot escape—even at the level of electrical signals routing through circuitry—the inevitable slippage that language itself produces.[7] Programming is itself subject to the very qualities that afford, in human speech, the untethering of buzzwords like AI from their material circumstances. In other words, despite the popular image of computers as material rationality, exemplified by the fictional engineer Charles Forbin's statement that the titular, sentient computer in the mid-century thriller film *Colossus: The Forbin Project* "deals in the exact meaning of words,"[8] there lurks an irrational, concrete kernel that forever renders the full reality of their operation inexpressible in abstract, semantic terms. In modeling themselves after the brain, however, CNNs move significantly beyond the inevitable introduction of bugs, unexpected behavior, and security threats attributable to the inexactitude of language. Rather, they verge on the philosophical since, for the German

Idealist thinker G. W. F. Hegel, the mind being able to comprehend itself is the ultimate goal of a dialectical process by which the real becomes rational and vice versa.[9]

The problem that presents itself with CNN AI is that computer-modeled "neurons" are "trained" to produce results in such a way that explaining how they are derived is almost out of the question, even with full insight into the way the algorithms that produce them generally function. This inexplicability makes them qualitatively distinct from determinate, yet computationally difficult or irreversible, processes like cryptocurrency mining, even as they threaten to take up even more computing power and electricity than these applications. From an informational perspective, neural-network outputs are highly entropic in every sense. They produce results typically unexpected of computational processes but do so unreliably, at times delivering the "desired" output and at others not, with limited opportunities for fine-grained refinement.

Many compelling AI applications are focused on the processing of audiovisual data, and several of the most influential papers in the field are likewise centered around image recognition. As is well known, the basic process for training such networks, which are not networks in the sense of computer networks but rather algorithmic models made up of interconnected variables that change in relation to each other, involves feeding a randomly initialized network a series of hand-labeled data, in this case, images. The system is then tested on what it has "learned" via a subsequent set of similar, but unlabeled, information. The results of this second, testing phase are repeatedly evaluated and cybernetically "back-propagated" through the layers of neurons that produced them, each time modifying the "weight" at which each neuron is "triggered" to either reinforce (in the event of a correct result on a test image) or reconfigure (in the event of an incorrect classification) the operativity of the network as a whole.[10] At some point this process is declared effective enough for a stable—that is, commodified—copy of the network to be put into production.

The combined output of an aggregate of individually trained, mostly independent processes is not easily explainable via the logic with which we are used to being able to describe computer programs, however inaccurate it might be. Cognitive scientists, including two of the developers of the back-propagation technique, have characterized NN-like techniques as operating at a level of consciousness akin to that of quanta in physics: "What PDP [parallel distributed processing] models do is describe the internal structure of the larger units, just as subatomic physics describes the internal structure of the atoms that form the constituents of larger units of chemical structure."[11] As with the state of quantum particles, this unknowability remains the case today, as contemporary researchers admit of the "current CNN model" that they "do not even know why and how it works essentially."[12] While engineers are able to identify individual "feature extractors" highlighting elements such as a specific type of curve, color, or texture that contribute to the recognition of, for instance, a specific object or written character, the ultimate classification is made only as the cumulative result of a large number of individual, often contradictory recognitions and is hardly reducible to formal, logical definitions.[13] Paradoxically, these erratic output generators are considered ideal for dealing with entropic input: "With enough computation and enough data, [machine] learning beats programming for complicated tasks that require the integration of many different, noisy cues."[14] It is precisely the tendency of handwriting, photography, and other indices of the real *not* to be perfect representations of some Platonic ideal but rather messy, complex impressions that CNNs are meant to address.

Thus, while at the level of machine code the operations of CNNs are as clear (and abstruse) as any other program, the myriad, miniscule relations between each individual element of a picture and each individual feature extractor are not easily comprehensible outside of the final result. It is this aspect of CNNs that has enabled the more-than-cottage industry of AI ethics based on, for example, verifying that applications geared toward identifying and classifying people and their behavior respond "correctly" to a full breadth of informationalized representations of human identities. It also raises more vexing questions about the relationship of these "unauthored" systems to existing sociopolitical institutions such as the legislature and judiciary. As one famous example of the latter goes, in an incident where a "self-driving" NN injures someone or damages something, who is answerable?[15] Moreover, how can such errors be said to be fixed when developers cannot really understand what led the network to make a given decision?[16]

Computer Power and Electronic Irrationality

I have so far given what amounts to a summary of the state, both technologically and socioculturally, of CNN-based AI. It is not a coincidence, however, that we have come to embrace this apparently radical break from previous notions of technological rationality. More to the point, it is no happenstance that we have come to do so *now*. In exploring why, we again re-encounter that very quality that slid away when we began to conceive of computation as the abstract, slippery play of signification inextricable from language and equally divorced from physical reality: the material world. One of the driving factors of recent AI hype is the looming end of Moore's law. The amount of computing power that can be deployed via the predominant, serial, single-processor, discrete silicon model enshrined conceptually since the mid-1940s with the EDVAC's von Neumann architecture has been a direct function of the increasingly microscopic scale at which semiconductors have been fabricated since the early 1960s, when this design became physically dependent on a silicon rather than vacuum-tube base.

Ironically, as Chun points out, the von Neumann architecture was originally conceived in terms of a symmetry with neurons, albeit a highly abstracted, cybernetic version of them, and John von Neumann himself initially considered NNs to be capable of the same type of exacting rationality attributed to the fictional Colossus mentioned previously.[17] As she describes, such formulations would not last, with hardware rapidly moving away from a notion of memory as executable sets of firing neurons and toward memory as semi-animate storage, with software in turn adapting to take user input during runtime to make up for difficulties proceeding mechanically from supposedly rigorous first principles.[18] By contrast, contemporary AI moves back away from being guided by users and even programmers. As she writes, "machine learning algorithms . . . reveal . . . that our computers are now more profound programmers than their human counterparts."[19]

While this profundity is arguably up for debate, the parallel-by-default nature of CNN AI nevertheless does present itself as a much-needed alternative to traditional software and hardware development. It promises to revitalize the computing industry just as it struggles with declining levels of growth in established paradigms. NN frameworks may be written in conventional programming languages, but their specific power is achieved by their subsequent self-configuration to process any type of data in parallel. Indeed, one major breakthrough in the widespread adoption of the technology was the 1,900% speed increase achieved with commodity hardware, at precisely the moment the number of clock cycles per second in standard central processing units began to stagnate, by offloading NN processing from CPUs to the GPUs packaged in add-on cards popular with gamers.[20] Continuing increases in GPU power, driven by optimization of a subset of the operations carried out by CPUs, led to a watershed moment in 2012 when a team, including one of the earlier proponents of back-propagation and parallel processing, leveraged such a system to win that year's ImageNet Large Scale Visual Recognition Challenge, a major automated image-identification competition.[21] Improved performance of their training process on newer hardware cited in a keynote by GPU maker NVIDIA's cofounder and chief executive Jensen Huang prompted the notion of "Huang's law" superseding Moore's, as though graphical processors operated independently of the physical constraints afflicting general-purpose computation.[22]

Despite these twenty-first-century hardware enhancements leading to broader adoption of the software technique, the relationship between NNs and parallel processing had already been understood for decades. As is well known, despite early cybernetic conceptions of NNs, AI as a practical subfield of computer science largely settled initially around classical paradigms like structured, functional, and imperative programming. This approach was pioneered by luminaries like Joseph Weizenbaum, whose well-known ELIZA program, in its DOCTOR role, simulated psychotherapy simply by parsing the sentences fed into it by a user and returning them as questions.[23] A key takeaway from this experience, Weizenbaum wrote, was "that extremely short exposures to a relatively simple computer program could induce powerful delusional thinking in quite normal people."[24] In contrast to the simpler, sequential programming methods used to create programs like ELIZA, more complex distributed models, long accepted as state of the art in cognitive science but less prominent in AI, "began to have something of a renaissance in computational circles" roughly contemporaneously with the publication of Weizenbaum's book.[25] The processing power to support them, however, remained lacking, as with the mid-1970s DARPA-funded speech recognition system HEARSAY, whose "architecture was too demanding for the available computational resources."[26] Established and largely serial programming paradigms thus remained dominant in practice.

At the same time it became clear that "logical" solutions, particularly those encoding a high-level set of either concepts or cases by which a computer could proceed in some area of AI, would not scale due to the complexity of real-world situations.[27] Questions of scale are really questions of performance, or, more bluntly, time. Can a computer process the necessary information quickly enough to put it to use, for example, to convincingly simulate a human being, as in the classic Turing test, or to swerve a car to avoid a collision?[28] After a point, and contra media theorist Vilém Flusser, the real, where special situations and informational noise abound, came to be understood as too complicated for the rote application of Platonic forms, necessitating a new foundation for AI.[29] By the mid-1980s, parallel processing had been established, at least in theory, as a solution to the "brittleness" of top-heavy rule-based, rather than inductive, systems: "In order to avoid a distinct rule for each situation . . . it is imperative that the system's response to any situation be mediated by the concurrent activation of a set of relevant rules."[30] Although theoretically possible in both, this adaptable, concurrent mediation is best utilized via an "'environment-oriented'" rather than "'language-oriented'" approach, one that "does not explicitly assign abstract symbols to" input, "nor does it explicitly search for and apply grammatical rules to such symbols."[31] Verb conjugation, for instance, can proceed via developing associative connections between neuron-like elements, feeding them pre-classified sets rather than delineating and processing fixed cases; this approach, researchers argued, demonstrates more similarities with linguistic development in children than rule-based models.[32] While such parallel systems do not necessarily have to be NNs, the definition of independent yet aggregable concurrent microscale information processors comports well with contemporary models of the brain.[33]

Conceptualizing the Brain

For philosopher Hubert L. Dreyfus, who covers much of this history and more in his prescient critique of AI, first published in 1972, revised in 1979, and updated in 1992 with the amended title *What Computers* Still *Can't Do*, the distinction between 1960s and '70s language-based models and the multilayer NNs of the mid-'80s is one between "Good Old-Fashioned AI (GOFAI)" and the cognitive-science "connectionists."[34] Dreyfus's original work offers an early argument, drawing upon phenomenology, against the logical, rules-based model of AI's compatibility with the workings of the mind due to both practical scaling issues and, more fundamentally, philosophical incommensurability. While, as he notes, this older paradigm continued to struggle along, in the updated edition he provides a startling parallel to describe the rapid resurgence of NN connectionism. "Like the dissolution of the Soviet Union," he writes, "the speed of collapse of the GOFAI research program has taken everyone, even those of us who expected it to happen sooner or later, by surprise."[35] This metaphor would seem to be somewhat common, as Hegelian philosopher Catherine Malabou, in her consideration of contemporary neuroscience, also makes use of it to describe the difference between a centralized, command-based model of the brain she associates with Cold War cybernetics and the dominance of distributed, allegedly stochastic and antihierarchical, methodologies in the post-Soviet age of digital networks and mediation.[36]

Connectionist approaches to AI, in contrast, have more or less congealed around this latter model via CNNs, providing further evidence of Malabou's critique that an ideological confluence exists among networks, cognitive neuroscience, and neoliberal capitalism. That such associations have come up repeatedly in critical scholarship on artificial intelligence should not be surprising, as they are already gestured toward in the technical literature. For example, in at least one cognitive-scientific text that is a key influence on parallel processing models, these correspondences go beyond the concept of "mental . . . or psychological Darwinism" that Malabou notes, as well as other homologies between neurons, computing, evolutionary theory, and an individualistic, egocentric capitalism identified by Chun,[37] and into explicitly economic metaphors. Individual processes akin to feature extractors are described as both "genetic" *and* tasked with "'making a profit'" such that "the classifier's strength is a kind of capital."[38] This description operates at a deeper level than the way that, as Malabou writes, "very often, the brain is analyzed as personal capital, constituted by a sum of abilities that each must 'invest optimally.'"[39] It moves, instead, into the component parts that comprise, for contemporary neuroscience, both the brain and its abilities. Here, each rule (later, neuron) is conceived as an individual actor in the network-as-market that must adapt not simply evolutionarily but specifically to self-valorization or perish.

Malabou describes this logic more broadly as one of "flexibility," a kind of accession to the demands of globalization such that "to be flexible is to receive a form or impression, to be able to fold oneself . . . to be docile, to not explode."[40] Indeed, this concept and language can be found not only in neuroscience and neoliberal self-help mantras. It is also present in arguments that parallel processing can help "capture the very flexibility and open endedness of cognition."[41] We are thus confronted not simply with an analogy between neuroscience and neoliberalism, but the technologically enabled invasion of capital into previously insulated spaces—those of state socialism and the cerebrum alike.

In his updated edition Dreyfus makes clear he is also unconvinced by the "environmental," neural approach, a conclusion he supports with reference to the obfuscation represented by intermediate layers, or feature extractors, described earlier. Drawing on research conducted with his brother Stuart, he notes that differently initialized networks adapt differently to the same input, shattering "the rationalist intuition that these explanatory features capture the essential structure of the domain."[42] Not only does this return us to the irrational kernel of NNs but, further, the move from the concrete to the abstract within it, such as it can be said to be taking place, paradoxically remains particular rather than universal. "If the net *had* abstracted a theory of the domain," Dreyfus points out, "it would have converged on the correct features and so would not have had to be further corrected" via back-propagation and so forth.[43] Whether in his real-world example, of a military AI distinguishing between a sunny day and a cloudy one instead of flagging the presence of camouflaged tanks due to the times at which its "true" and "false" training photos were taken, or the more recent story arc of the satirical television show *Silicon Valley*, where an AI tasked with optimizing a communications network devotes itself to breaking state-of-the-art encryption to better compress the network's plaintext messages, in the gap between a definitive NN for a given domain and the inherent variation of individual networks trained on the same data, there lies the infamous tendency of NNs to adapt themselves in ways other than those intended by their curators.[44] Due to the opacity of the intermediate layers, such adaptation reveals itself, as AI ethicists well know, only in the case of "faulty" responses to new stimuli.

Rather than dialectically creating abstractions from the concrete messiness of the real world, then, the NN approach reinforces this uncertainty. This further places it in line with a neoliberal tendency to emphasize the complexity of a problem in order to offer an equally complicated solution as the only viable response. Partially for this reason, and despite Dreyfus's acknowledgment that, unlike GOFAI, NNs operate "in parallel, thus avoiding the bottleneck of serial processing,"[45] they did not immediately take off as a practical field in the 1980s. Even with additional innovations such as reinforced learning, whereby systems can derive reward and punishment metrics directly from their actions in an environment, Dreyfus concluded "that our current ignorance concerning the brain and practical limitations on computer memory size make it highly unlikely that there will be substantial progress toward this kind of brain-inspired AI in the foreseeable future."[46] The capacity of solid-state memory, the dominant material form of RAM over the past fifty years, is, like processor performance (which NNs leverage to keep memory requirements manageable), determined by semiconductor density and therefore subject to Moore's law. We thus return to the question of why NNs were eventually widely embraced and specifically why they were *not* broadly adopted until after the turn of the century.

Before addressing this question directly, however, it is necessary to explicitly disambiguate two thus far implicitly connected concepts. The advantages of contemporary NNs are both practical, minimizing the chokepoints of sequentiality, and theoretical, resembling more closely present biological templates of the brain. These strengths do not necessarily have to coexist, though it is clearly beneficial when they do. We have seen already that NNs are only one type of parallel process. Even a "language-based" system could operate in parallel as long as its individual evaluators could work largely independent of one another and thus be allocated concurrently to multiple processors. Although presenting difficulties for a paradigm typically entrenched in serial concepts, such processes can nevertheless be written by hand, minimizing (but never fully eliminating) moments where the distributed processing streams come together to form a chokepoint in accordance with Amdahl's law, the principle established just two years after Moore's characterizing the extent to which any informational procedure is ultimately parallelizable.[47]

Similarly, while in the neuroscientific and NN views the brain is treated as a parallel information processing "machine," there is no reason a single processor system could not model it, provided its developers were willing to take the requisite performance hit. That is to say, a discrete, electronic serial computer could emulate a configuration of artificial neurons in parallel; it just would not be able to do so as quickly as an actually parallel one[48] (or, as we shall see, a brain—accepting here, but only for the sake of argument, that the brain is indeed a kind of computer). Gene Amdahl's original formulation even concludes by hypothesizing that a custom-built serial processor would *outperform* the parallel associative processors of the late 1960s on "pattern recognition" tasks, distant precursors of those routinely undertaken by CNNs.[49] With modern technology, where parallel processing has become more effective as single-threaded performance has stagnated, multiple serial processing cycles would be required to simulate a single parallel processing cycle. This is not unlike the way that a single instruction on a complex instruction set computer (CISC) can take up more cycles than those in nimbler reduced instruction set (RISC) architectures. The historically higher rate of increase in serial clock speed can therefore account for the inability of specialized NN processors to take hold until recently.

Neural Networks and Moore's Law

A key 1991 paper, for example, describes a specialized, US Army–funded hybrid serial–parallel digital–analog chip designed with CNN applications in mind because "the computational requirements, data rates, and size of neural network classifiers severely limit the throughput that can be obtained with networks implemented on sequential general-purpose computers."[50] The chip was not meant for training NNs but rather only for production usage—that is, the deployment of a previously trained NN for the purpose of on-the-spot classification. The tested domain, specifically, was handwritten character recognition, where results indicated a fiftyfold speed increase of the hybrid chip in comparison to a commodity digital signal processor alone. Due to Amdahl's law, this number is itself constrained in comparison to the performance of internal on-chip operations that are fully parallel.[51] The researchers thus concluded that "the availability of fast special-purpose hardware for large applications permits exploration of new neural network algorithms and problems of a scale that would not be feasible with conventional processors."[52]

Yet, one of the breakthrough papers on AI handwriting recognition, describing a CNN dubbed LeNet-5, published nearly seven years later and cowritten by Yann LeCun, one of the authors of the previous study and a major figure in AI, relied instead upon computing breakthroughs in commodity hardware. Citing the earlier report, the latter group writes, "single chip mixed analog–digital implementations of LeNet-5's predecessors have been shown to operate at speeds in excess of 1000 characters per second. However, the rapid progress of mainstream computer technology renders those exotic technologies quickly obsolete."[53] There is a simple explanation for this: Moore's law dictates a doubling of semiconductor density, and thus (usually) processing power, roughly every twenty-four months. For comparison, while the 1991 chip used somewhat more than 180,000 transistors manufactured at a scale of 0.9 µm, or 900 nm, a state-of-the art serial Intel 486SX CPU from the same period was made with 1.4 million transistors at a process ranging from 1–0.8 µm.[54] Intel's 1998 Pentium II processors were made with over five times as many transistors, at roughly a quarter of the scale.[55] Even accounting for the custom chip's performance differential due to parallelization and analog–digital hybridity, then, researchers could passively rely on Moore's law for even greater gains, all without having to develop their own hardware.[56] This is particularly true given not just raw scalar metrics but additional, qualitative processing capabilities. These begin, for Intel, with the 486DX's incorporation of a previously optional math coprocessor—the LeNet-5 authors reference "low-cost machines with fast arithmetic units"—and extend, as they note, to the company's MMX instruction set, introduced in 1997, rendering "the success of specialized hardware hypothetical at best."[57] The same trend that kept classical programming paradigms afloat was also working to the benefit of CNN researchers. Their parallel models of the brain could continue to advance on the back of improvements in serial processing power.

Stagnation in clock speeds around 2004 threatened to put an end to this kind of growth. However, in addition to the shift toward a more inherently compatible parallel CPU architecture, a second development around the same time helped to exponentially accelerate it. Graphical processing units had become increasingly present in commodity PCs, particularly those oriented toward the gaming and 3D graphics markets. GPUs are typically not (yet) part of the CPU or motherboard and are made by separate companies such as 3dfx, ATI (subsequently acquired by Intel competitor AMD), and NVIDIA (which eventually absorbed 3dfx). While neither as computationally versatile as CPUs nor as optimized as FPGAs or ASICs, they excel at the parallel computation of a more limited set of operations crucial to CNNs, particularly the one the 1991 custom chip was designed to optimize: matrix multiplication.[58]

Put as simply as possible, the basic process of CNNs is to "convolve" two matrices: one representing the neurons, essentially a two-dimensional array consisting of a trained set of weighted values, and the other representing an equally sized portion of the input. For example, a 4×4-pixel feature extractor may be "moved" across a 16×16-pixel black-and-white image. In such a scenario, each pixel will be assigned a value, for example, 1 if it is black and 0 if white. The upper-left value then will be multiplied by the weighted value associated with the corresponding neuron and added to the sum of the products of each pixel in its corresponding row and each neuron in its corresponding column. This process is repeated down and across each matrix, resulting in the matrix product of the values of the input data with those of the extractor. This convolution is then run through a sigmoid function for nonlinear normalization, the position of the extractor incremented (typically by only one or two pixels to ensure overlap, which helps the network identify the same features regardless of their position within the input matrix), and the operation conducted again on the new data. These results are then used as input for a higher-order layer, along with those of many other extractors. The process will essentially be repeated until the final layer, which generates a classification or other decision, is reached. Individual extractors are relatively independent and can thus be parallelized.[59] More recently, multiple GPUs can themselves be configured to run in parallel. Combined with the built-in strengths of GPUs, whose graphical functions can be used to quickly perform matrix multiplication via vertex shaders and sigmoid normalization via pixel shaders, this architecture made large-scale CNNs practicable on commodity hardware.[60]

Toward a Materialism of AI

To return now from the concrete to the abstract in what is hopefully a more comprehensible move than those routinely made by CNNs, this attempt to historicize developments in the past forty years of NN research, software, and hardware has also been an effort to reinject the material grounds of computation into a discussion usually dominated by other considerations. Computing performance, both in its exponential heyday and the post-2004 lean years, significantly shaped—one is tempted to say, determined—the practical viability of these theoretical techniques. As the need for parallel processing paradigms became more urgent, Herb Sutter told programmers, "there are two ways to deal with this sea change. . . . One is to redesign your applications for concurrency. . . . The other is to be frugal, by writing code that is more efficient and less wasteful."[61] But NNs represent a third approach: offloading concurrency onto neural frameworks. Such a paradigm was very much in demand, as the National Research Council report on the faltering of Moore's law, written before CNNs had risen to the prominence they now enjoy, makes clear: "The shift to explicitly parallel hardware will fail unless there is a concomitant shift to useful programming models for parallel hardware."[62] As Sutter's article charts, it was only once the CPU clock ceiling had been reached that parallelization became widely viable. This is even more true for parallel-by-default NNs, despite the mainstream existence of 3D accelerator cards for almost a decade prior and multiprocessor systems in niche areas like servers for even longer. Similarly, it was only once GPU hardware became widely adopted for CNN tasks that a broad consensus on the importance of AI, including the development of more specialized hardware for the marketplace, began to congeal.[63]

Notably, many contemporary reports, even when they acknowledge these material developments, choose instead to emphasize alternative determinants. As a white paper by former Google AI policy executive and MIT Media Lab Research Affiliate Tim Hwang notes, "While a common trope in both the popular press and scholarly literature is to highlight the massive increase in computational power that has enabled the recent breakthroughs in machine learning, the analysis frequently goes no further than this observation around magnitude."[64] Despite mentioning them, both the LeNet-5 developers and the 2012 ILSVRC winners, for example, stress the availability of collated and/or labeled datasets and algorithmic, software-based innovations over and above improvements in hardware.[65] As with FPGAs and ASICs more broadly, the AI software landscape is rapidly changing, in a mutually determining if not dialectical relationship with advanced hardware, to the point where algorithmic optimization may end up eliminating the need for matrix multiplication (though not parallelization or specialized hardware).[66] At the same time, much thinking in the space, including the 2012 paper, falls back on the same comfortable notion that drove conventional software engineering for nearly forty years: "All of our experiments suggest that our results can be improved simply by waiting for faster GPUs and bigger datasets to become available."[67] Even Hwang suggests something similar about not only GPUs but the custom silicon now emerging.[68]

Yet, as Sutter writes, "'Oh, performance doesn't matter so much, computers just keep getting faster' has always been a naïve statement to be viewed with suspicion, and for the near future it will almost always be simply wrong."[69] Despite the enhancements represented by specialized hardware and their ideological obfuscation as a consequence of "Huang's law," the physical constraints affecting Moore's law hold there as well, and for potentially much longer than the "near future." There is only so much room for software- and data-based improvements before the same material factors bedeviling CPUs affect GPUs and ASIC "AI processors" as well.[70] Moreover, as Amdahl's law indicates, there is only so much of any task that can be parallelized.

Beyond Moore's law, yet hardly unrelated to it, there remains a further, even more fundamental issue with attempts to model the brain in discrete hardware, silicon-based or otherwise. This limitation manifests itself, albeit obliquely, if we return once more to Friedrich Kittler's essay "There Is No Software." There, as we have already seen, Kittler ultimately concludes that a completely different paradigm of computing may eventually appear, "under evolutionary disguises or not," called into being by the "the very isolation between digital or discrete elements" that "accounts for a drawback in connectivity."[71] In doing so, he again draws upon Michael Conrad's description of these limitations in detail. Conrad establishes that discrete machines are constrained in the number of interconnected elements they can make use of in comparison to organic, analog "computers" like the brain.[72] The closest approximation, and thus in a sense the most efficient use of discrete computing power, he concludes, also deploying an economic metaphor, is modeling the connectionism of the brain. "*Representing evolution-enhancing features in structurally programmable systems,*" Conrad writes, "*can finance itself to the extent that it increases the effectiveness with which computing resources are used for problem solving.*"[73] Specifically, Conrad proposes a NN-like configuration composed of cellular automata running on a massively parallel computer designed to "evolve" its software on the basis of random permutation and fitness selection, including the use of specialized hardware designed for this purpose.[74] While such a system would never be equivalent in processing power to the brain, it could be the closest that a discrete, structurally programmable computer could manage.

This theoretical analysis is demonstrated in practice by contemporaneous custom hardware like the 1991 chip described earlier. Despite being a hybrid of discrete and continuous modes of mediation, this hardware implementation prunes over 99% of its global connections in favor of localized ones to maximize efficiency.[75] Such shortcomings in relation to the actual capabilities of the brain become even more glaring when we return to the fully discrete computation upon which modern parallel multi-core CPUs, GPUs, and machine-learning accelerators rely. All these AI implementations are foundationally enabled by paradoxes. Not only do they attempt to model the analog, electrochemical operations of the brain (suggestive as the neuronal fissure is of discreteness)[76] via digital, binary signals, but, moreover, the quantitative operations of these processors are themselves interpreted qualitatively, in terms of a subjective accuracy and time to completion. Neural networks are usually judged on criteria like whether a facial recognition program can identify images of people to a developer's satisfaction within a timeframe established by human temporality, or aspects of the images or text they produce. Indeed, they can be interpreted *only* qualitatively, since "traditional programming, or prescriptive control over the rule obeyed, is given up."[77] Thus while the question posed by Malabou, *What Should We Do with Our Brain?*, will be quantitatively and, therefore, qualitatively different for CNN AIs than for humans, the latter are nevertheless in the process of trading rationality for short-term performance gains under the existing hardware paradigm threatened by the end of Moore's law.

The dialectical counterpart of this contemporary embrace of CNN AI is an education system focused on "flexibility," on practical skills and "entrepreneurship" over speculative thought.[78] As noted scholar of automation David F. Noble's turn-of-the-century analysis observes, "the major change to befall the universities over the last two decades has been the identification of the campus as a significant site of capital accumulation."[79] He identifies this trend first in the increasing prominence of practices like licensing research patents out to industry in the wake of the early 1970s neoliberal turn and subsequently in the development of similarly licensable digital course materials. Meanwhile, Catherine Liu writes, today's "global education leaders . . . believe that education should produce entrepreneurs who in turn produce economic growth while they advocate for information and communication technologies."[80] The emphasis on growth, in other words, has shifted toward so-called job creators and away from those who would fill those jobs—that is, those providing the surplus value upon which growth is predicated. It is instead supported, or imagined to be, by technological innovation.

Ironically, this drive operates as something of a feedback loop as, on the one hand, for Liu, "the instrumentalization of all aspects of education appears increasingly attractive . . . to the economically insecure."[81] On the other, as Gary Hall has convincingly argued, the ongoing neoliberalization of the university along the lines of digital "sharing" platforms like Uber and Airbnb means that

> academics are likely to prefer to run courses in subjects that are perceived by student debtors-as-consumers as having the potential to help them gain a "good" job with a decent salary. They will thus be involved mainly in producing the type of unthreatening, lower-level, vocational "workers" that are needed by postwelfare capitalism (and which the current push on the part of many governments toward an "employability agenda" for much of higher education seems determined to generate) rather than the kind of educated public citizens or creative critical thinkers who are capable of maintaining some control over their own work and futures (and who therefore might not be quite so focused on the maximization of production and profit).[82]

Thus, paradoxically, while being encouraged as entrepreneurs, students also internalize the demand that they enter the workforce trained on the tools du jour rather than the concepts behind them, as what Noble calls "'just-in-time' educated employees."[83] The focus is then on Adobe Premiere, not century-old Eisensteinian montage, the latest fly-by-night web framework over fundamental insights into procedure, JavaScript over Lisp. This is equally true of the faculty, who "are drawn into a production process designed for the efficient creation of instructional commodities" in which "the technology is being deployed by management primarily to discipline, deskill, and displace labor,"[84] even if the specific efforts to automate this labor-intensive work have changed over the past two decades.

Indeed, the relationship between labor and automation in education can be increasingly characterized in terms of a tension between the renewed interest at both research and teaching levels for covering AI on the one hand and the fear that neural-net-based tools like ChatGPT will obviate the higher-order thinking traditionally developed through writing-intensive courses on the other. It thus remains the case that, while "some skeptical faculty insist that what they do cannot possibly be automated," as Noble writes, "it will be automated anyway, whatever the loss in educational quality."[85] The qualitative difference the rise of AI poses is that this automation can now be put to use by educators' students as much as their administrators. Is it surprising then, that, as Hall notes, "universities are themselves shifting much of their attention and resources away from teaching and research to focus on management, marketing, and income generation"?[86] Malabou's hopes for a critical theory of neuroscience notwithstanding, the vogue for CNNs not only does not get us closer to a reconciliation between mind and matter, it forestalls it in the name of bolstering machinic efficiency while fostering merely quotidian capacities for thought and action in people.

In this sense, Malabou's description, just before the GPU revolution, "that we live in a period of 'weak' Artificial Intelligence" remains accurate despite the closer relationship between parallel computing architectures and the contemporary models of the brain she poses against cybernetic paradigms of programmability.[87] For all their advantages, when CNNs provide associative streaming suggestions, identify friends in photos, or attempt to navigate real-world landscapes—and often even when they respond to our chat prompts—there is little risk of confusing them with people. Likewise, touted applications for AI like image enhancement can produce mixed results. For every fan-made upscaling of a classic game like *Final Fantasy VII*, there is a headache-inducing "remaster" of *L'Arrivée d'un train en gare de La Ciotat*, whose flickering "improvements" are taken as state-of-the-art only insofar as someone unwittingly compares them to a low-resolution copy.[88] Powerful delusional thinking, indeed. Unlike viewers of the Lumière Brothers' film, we are witnessing not the birth of a new medium but the slow death of a current one. Even the most interesting developments, like image- and text-generating frameworks, are materially bound from making associations to the extent a human can, rendering these simulations a poor approximation of their referent, whatever other uses they may have. Indeed, in a logical bottom to recursion that gestures at these larger physical limitations, AI loses effectiveness when trained on the products of previous-generation AI rather than human-made content.[89]

Such limits put into perspective the hullabaloo around the now-obsolete GPT-2 model, whose developers initially and sensationally refused—citing and thus reinforcing the contemporary liberal mania over disinformation—to release it in full. Such systems cannot approximate the level of connection in the brain without a massive investment of resources that would render them ineffective for the very purpose of real-time intervention in online discourse.[90] If we nevertheless assume that they can do so successfully, we demonstrate our belief not in artificial intelligence but rather that most human receivers of information are stupid. The logistical challenges posed by this differential between cognition and its algorithmic approximation are substantial. Precisely because of the flaw highlighted by Dreyfus—that training NNs does not result in a single best solution to a given problem—as previously mentioned, the energy consumption necessary for AI models to even approach humanlike results threatens to dwarf that of the proof-of-work cryptocurrencies discussed in the previous chapter. Regardless of what one thinks of Bitcoin and its ilk, directing massive amounts of computing power toward AI would result in a much less quantifiable set of outcomes. From these, the wealthiest tech companies would most stand to gain (if there is anything to gain). As such, the nascent AI industry is already experiencing one of the most prominent limitations presented by the end of Moore's law: its ambitions far outstrip the availability of computing power to fulfill them.

There is, however, one affinity with the classical model of AI going back to Turing that is not simply overlooked but seemingly intentionally obfuscated. Even at their most neuroscientific, contemporary CNNs are undeniably symbol processing machines first and "neural networks," if at all, a distant second. This is a consequence of the paradox outlined above: Any model of the brain is just that, a model, and in the present case, one implemented on hardware that is discrete and electronic. Indeed, this is a paradox even for Malabou, who argues simultaneously that the brain itself is not comparable to the programmed, centralized, cybernetic image of computers, "that is, physico-mathematical structures endowed with the property of manipulating symbols," but also that the brain "represents itself to itself" via preconscious systems such that "neuronal activity constitutes the prototypical form of symbolic activity."[91] Like NNs, the workings of the actual brain remain in many ways a mystery.[92] Unlike the brain, however, in NN AIs, this "prototypical form" is itself built quite unambiguously on top of not just hardware but a symbolic, binary order. The consequences of this, as I have tried to show, are legion, but there remains one that deserves consideration here.

As noted, this process, at least on contemporary hardware, is entirely discrete. In other words, input must be digitized and turned into an idealized mathematical function to be analyzed. Neuronal feature extractors are digital, too. They are functions convolved across the input, composed of and producing discrete values that ultimately cascade into a decision expressible as one of a collection of possibilities. Such evaluations are measurable as information in Claude Shannon's foundational definition.[93] Despite implications otherwise,[94] then, every step of this process, from input to output, takes place at the level of symbol manipulation plain and simple. The power of NNs may not be solely attributable to idealized forms in Flusser's sense, but we are no freer from the constraints of language than those of the material conditions of computation. Yet, in adopting CNNs, we behave as though we are unfettered by both.

For Chun, drawing on Marxist psychoanalytic theorist Slavoj Žižek, this mode of behavior is synonymous with the way software operates as ideology. Žižek, himself building on philosopher Peter Sloterdijk, reformulates ideology from its orthodox Marxist roots as an economically determined intellectual mistake into a form of *behavior* contrary to our avowed understanding.[95] To this, Chun adds that software simultaneously represents both the process of demystification, in that it enables us to make previously hidden connections, and ideological behavior, in the sense that, mediated by programming languages and interfaces, we treat our computers differently from how we (on some level) know that they actually work.[96] A material analysis of CNNs updates this argument to account for the practical shift "toward less strictly 'programmable' systems" she describes as (at the time) largely theoretical.[97] More importantly, it allows us to factor in physical constraints alongside the effects of symbo-logical dissemination. It would be impossible to trade speed for ease-of-use with high-level programming languages, swap analog computers for digital ones (and vice versa) on the basis of performance, or hint that cybernetic models of neurons, while still a form of *logos*, may be less so than human-written code[98] if we did not have some sense of how well computers operated, some measure of their efficacy in the real. Any practical replacement of the programming paradigm with that of NNs would never be entertained unless it resulted in an increase in operativity. With NNs, instead of an abstract and erroneous understanding of a fetishized source code (though source is certainly still there, in their frameworks), we paradoxically abandon some abstractions, false or otherwise, to fetishize the more concrete, multifariously unknowable configurations of weighted simulated "neurons."[99] This begs the question of determinism, which in turn returns us to the earlier analogy between neural networks and quantum physics.

In a short but powerful essay, the Italian philosopher Giorgio Agamben makes a seemingly abrupt departure from the deconstructive approaches that also inform readings of the ineffable slippages in symbolic programming to put forth a startling claim against the entire paradigm of quantum physics. Investigating the late work and mysterious disappearance of the mid-century Italian theoretical physicist Ettore Majorana and drawing on radical philosopher Simone Weil, Agamben levels an explosive charge: Quantum physicists abdicated their scientific and philosophical responsibility to adjudicate the workings of the real. Quantum physics famously denies determinism full stop. The most fundamental workings of the universe proceed by chance and are only recuperated into predictable structures at higher levels, thus the analogy made earlier with neuronal operations by way of explaining the "flexibility and open endedness" of thought, or, put another way, their basic incomprehensibility. Following Weil, Agamben writes, "As soon as we assume that the real state of a system is in itself unknowable, statistical models become essential and cannot but replace reality."[100] Such probabilistic projections, Agamben suggests, may be useful, as Majorana concluded, for the social-scientific "art of government," but not for the point at which science and philosophy connect: the project of positing how and why what is is what it is.[101] Government, of course, is at the etymological root of cybernetics as well as the conceptual basis for the thermodynamic engineering from which it and information theory derive their core concepts such as entropy, homeostasis, and the minimization of operational friction.[102] As we have seen, these notions have their own lineage in artificial intelligence via its origins in cybernetics and recourse to postwar control principles, not to mention the ongoing neuroscientific conception of low-level brain functions as facilitating homeostasis in the organic body. Thus, like the contention examined in the previous chapter that we have entered a postthermodynamic stage of capitalism, Malabou's claim that "the cybernetic metaphor" of the brain "has . . . had its day"—that we now inhabit a paradigm predicated instead upon a primitive, yet direct understanding of the brain simply via the addition of the capacity for change, or plasticity, to this metaphor—would seem premature.[103] The brain, whether simulated in silicon or conceptualized by scientists, may no longer be centralized, but it remains a collection of *cerveau*-mechanisms.

Regardless of what we think of quantum physics—and, following Agamben, I will note in passing that the existence of orderly higher-level physical structures begs the question of an as-yet-undetermined determination—none of the enigmas there apply to CNNs. The twenty-first-century trend toward AI, as with every recent development in computing, is increasingly determined by an encroaching diminishment in the rate of hardware performance improvement. In another essay, Kittler frames the distinction between hand-written code and CNN-based "learning" as a choice, much like Agamben, between the hard work of uncovering reality and willful ignorance:

> The so-called 'hidden layers' in today's neuronal networks present a good, if still trifling, example of how far computing procedures can stray from their design engineers, even if everything works out well in the end. Thus, either we write code that in the manner of natural constants reveals the determinations of the matter itself, but at the same time pay the price of millions of lines of code and billions of dollars for digital hardware; or else we leave the task up to machines that derive code from their own environment, although we then cannot read—that is to say: articulate—this code.[104]

Oddly, for Kittler this places AI and humans in the same situation vis-à-vis language rather than illustrating their material differences. Yet our rush to abandon causal explication here can only do so much to paper them over. Any "emergent properties of the microstructure"[105] of simulated neurons must still take place on top of the physical basis responsible for computing their interactions, namely discretized silicon semiconductors. While we may wish to draw a distinction between abstract, rule-based AI paradigms, which "grew simply by incorporating more data and operating faster," and NN approaches that facilitate embodied learning and development,[106] we can only do so if we misrecognize the lower $O(n)$ curve of the latter as reflective of the affective gap between stimulus and response in a way the higher curve of the former is somehow not. The push toward CNNs is, instead, conditioned—at least in large part—by the approaching physical ceiling to Moore's law. There is little mystery here.

These material technological conditions are mirrored by the more classically understood material determinants of political economy. Under the contemporary capitalist conjuncture, as new rounds of AI-based automation promise to take new rafts of tasks away from human intervention, working people are—once again—not freed to enjoy more leisure but increasingly reduced, as in education, to ever more menial tasks.[107] As Andrew Goffey notes, "Debates about the deskilling resulting from . . . the role of information technology in shifting the composition of the workforce only make sense to the extent that machine intelligence is understood as a possible substitution for human intelligence," but "that machines can replace humans tells us nothing special about intelligence, particularly if this is as part of an economy that, in its entropic repetition of the eternally self-same, generally produces stupidity rather than intelligence."[108] Not least of the dumb tasks human workers are presently subject to is classifying training data for CNNs via below-minimum-wage "micropayment" sites like Amazon's Mechanical Turk.[109]

In addition to automation, the larger geopolitical and monopolistic uncertainties of AI hardware development drive Hwang's concluding call for "governance efforts focused on shaping the direction and availability of computational power."[110] While countries like the US and China are headed for a competition on this front, since computing performance will not keep improving indefinitely, no governor can maintain its smooth functioning for long. Machines will of course continue their work after peak efficiency has been reached—quite possibly, in the case of AI, guzzling ever-greater quantities of energy. Yet even if the welfare state had not been gutted by fifty years of neoliberalism, there would be no social heat sink large enough to stave off the entropy generated by people's increasingly repetitive labors.

At the same time, real, positive change cannot be accomplished solely via greater awareness of the creative potential implicit in the distributed, plastic brain. Nor can it be addressed by an AI ethics that, in vigilantly overseeing the output of systems we no longer program as much as nudge, replaces politics, as Alain Badiou notes of ethics more broadly, with "the sympathetic and indignant judgment of the spectator of the circumstances."[111] It can only be achieved by synthesizing individual, concrete experiences of diminishing labor conditions into a broader mass movement, by sublating the specific histories through and against which, for Malabou, our separate brains have been formed into not simply "a new meaning of history,"[112] but History itself. History presents a limit to this increasing, and increasingly precarious, antinomy between the particular and the general because, as W. Daniel Hillis writes, "the only problems we don't know how to solve efficiently on a parallel computer are those for which the growing dimension of the problem is analogous to the passage of time."[113] History, to update Fredric Jameson, is not only "what hurts"; it is also and dialectically what it hurts to compute.[114] Otherwise, we would have no hope of transcending it.

The original Mechanical Turk, really a man situated underneath a chessboard making moves via a puppet passed off as an automaton, serves as a key figure for the mid-century Marxist theorist Walter Benjamin's "Theses on the Philosophy of History." For Benjamin, the puppet was analogizable to a revolutionary approach to history, while the man was meant to represent its hidden theological driver.[115] Caught between the fascist Spanish border and the Nazi army making its way through France, Benjamin took his own life soon after writing them. While subsequently celebrated, the theses were an act of desperation if not an outright appeal to divine intervention. In this they were enabled by his prior collation of fragments, "On the Theory of Knowledge, Theory of Progress," under the rubric "Convolute N," some of which were reused in the later text.[116] While, as the translators of the English edition of *The Arcades Project* make clear, the sense of "convolute" evoked here is the colloquial one, the effect remains the same as what we see in CNN hype: an inchoate theory of knowledge ultimately transformed into an appeal to superhuman intelligence. A fatalistic resurgence of interest in Benjamin's outlook, as Jameson has noted, coincided with the seeming triumph of global capital surrounding the events of 1989.[117] Today's workers, by contrast, do not need to wait for God. While time is running out for the logic by which they are governed, it is just beginning for them.

5 Winter in the Valley: The Accelerating Dynamics of Stagnation

The conceptual heart of the system of governance responsible for the management of computing power and, increasingly, workers, is, of course, Silicon Valley. This is perhaps best summed up by the phrase adorning many MacBooks and iPhones: "Designed by Apple in California. Assembled in China."[1] If the promises and pitfalls of Moore's law are registered anywhere outside of computing proper, it is here, in the region named after the semiconductor boom brought on in no small part by Moore himself. As previously noted, having long ago offshored most of its manufacturing capacity, the Valley has taken on an increasingly abstract role in the technology industry. While hardware and even transistor design still happen there, its torchbearers have shifted from electrical engineering companies first to software- and then service-oriented firms. These transitions have situated the region and its place in the cultural imagination at an increasing reserve from the physical fundamentals of computing that originally established its reputation. Resultingly, the impact of 100% utilization shows up not only in the material factors examined so far but also in the superstructural elements of tech culture built on top of them.

As with other cultural formations, these elements both mask and reflect ongoing problems with the perpetuation of the economic conditions enabling them. These are not simply physical limitations to semiconductor density; they are also barriers to labor, particularly in the so-called knowledge economy writ large. This latter paradigm is itself manifesting signs of distress as computing power and financial growth rates alike show indications of stagnation. Various ideological perspectives on unemployment and/or the future of labor have therefore come to prominence of late, influenced—knowingly or not—by the West Coast tech industry.

What these approaches do not consider, however, is the limited viability of Moore's law going forward. Indeed, in many cases, they assume the opposite: that computing power will continue to improve as it has done over the past sixty years. However, as we have already seen, technology and human labor are deeply intertwined, particularly in capitalist economies. Any proper reconsideration of our global digital society will therefore necessarily have to address our own relationship to computing power, and its relation to us.

Built on Sand

The shift from commodity products to service-based offerings over the last decade stems from the greater San Francisco Bay Area but extends into the way consumers experience technology worldwide. The Web 2.0, smartphone, and always-on cloud paradigms pioneered in the Valley had the effect of placing more people in connection not merely with computing—isolated instances of software and hardware—but, first via the internet and subsequently through its platformization, with tech companies themselves.[2] A milieu of 24/7 connectivity, walled-garden app stores, big data, machine learning, and monopolization meant that technology companies gained control over what individual users did with their hardware on a day-by-day basis in a way nearly unthinkable during the commodity era. In truth, the resulting cultural skepticism can be seen as an expanded, dialectical return to an earlier mode of criticism of and in the region in the 1970s and early '80s.

At that time, the Valley was more in thrall to the military-industrial complex and traditional corporate interests than consumer hardware, software, or services. The more economically oriented components of 1960s radicalism still held some sway over the famously liberal politics of the cities to its north and east. Since then and until recently, however, the focus on the straightforwardly exploitative and largely conservative nature of the high technology industry substantially diminished. Critiques in activist pamphlets, labor-oriented publications like *Processed World*, and literary explorations like Vikram Seth's *The Golden Gate* gradually gave way to movements like Critical Mass bike demonstrations, heartily adopted by the employees of new media and dot-com companies, magazines celebrating the new ethos they represented like *Mondo 2000*, *bOING bOING*, and

WIRED, and awkward attempts to represent its culture in films like *The Center of the World*.[3]

During this interregnum, the growing strength and ubiquity of computing power colonized all kinds of media production while being widely hailed as a democratizing force. A paradoxical development, however, took place in the early 2000s with the rise of Web 2.0. The first wave of the web, corresponding to the deregulation of the internet, had promised a leveling (though, crucially, not a socialization) of access to the mediatic means of production. This was famously expressed by one of the founders of the cyberlibertarian Electronic Frontier Foundation, John Perry Barlow.[4] Yet the proliferation of voices he celebrated was still limited to those with a relatively high level of technical knowledge. Web 2.0 addressed this problem, counterintuitively, by centralizing the operations necessary to put content online. It did so by embedding technologies like Asynchronous JavaScript and XML into web pages. In this way, for example, users could post and read new comments on a page without reloading it, making for a much more user-friendly approach to producing, rather than merely consuming, online information.[5] However, as Ron Eglash wrote not long after, "In terms of individual use [Web 2.0] is a move toward democratization through lay access, but in terms of business ownership it is a move toward monopolization."[6] The catch was that such user-generated data was stored not on user-controlled devices or webspace but rather on the servers associated with these pages.

Web 2.0 was the buzzword of the post-dot-com boom. It was the emblem under which much of the current generation of tech behemoths became household names, especially in the social media space. It was also one flown by slightly older companies such as Google, which pioneered the use of these technologies to dramatically expand beyond its core web search competency, for example, via its flagship web application Gmail.[7] Web 2.0 allowed everyday people to easily share their thoughts with nothing more than the web browser they already used. Yet it also drove them toward the increasingly narrow range of websites with the budget, talent, and infrastructure to make this experience seamless, enabling their users to easily answer the question famously posed by Facebook: "What's on your mind?" The wizards may still have been staying up late,[8] but their doings were increasingly dictated by mass-market concerns. Now more overworked than mysterious, the problem of a technically literate class was "solved" in a way that no longer elevated it but rather enshrined it in the middle layers of a small number of highly capitalized sites.

The initial launches of Gmail and Facebook and the broader adoption of the term Web 2.0 all took place in 2004.[9] The timing of this development thus corresponds not only with the economic recovery from the more mercurial dot-com boom but also with the onset of processor clock speed stagnation. The shift to massive social or participatory platforms with custom-developed backends dramatically increased the significance of "the cloud," where server farms, distributed databases, load balancers, and other parallelism-friendly architectures were already the norm.[10] On the front end, the transformation of the web into a technology for serving rich applications was exacerbated by the introduction of smartphones and eventually the app-based ecosystem they adopted. In this paradigm, preferential access to many websites is wrapped in a processor-native skin for increased performance, usability, and control. Such portable devices, by virtue of less flexible software and weaker hardware in comparison to full-fledged personal computers, further reinforced the importance of server-based solutions. Moreover, these apps are themselves typically vetted by the major smartphone developers and distributed through their proprietary storefronts. These developments consolidated corporate influence over the internet and guaranteed Apple, teetering on the edge of irrelevance throughout most of the dot-com era, a place in the pantheon of twenty-first-century tech giants.

The post–Web 2.0 technology industry made a reality out of the vision of Silicon Valley, and the Bay Area more broadly, that the first internet wave only foreshadowed. Complaints about gentrification, a major issue in San Francisco since the end of an early 1990s real-estate downturn, began to take on a new tenor as younger generations of Valley employees increasingly refused the trade-off their elders had been asked to make between a career in technology and city living. The subsequent introduction by firms like Apple and Google of private bus systems for shuttling urban dwellers to and from their jobs on the peninsula heightened the sense that not only San Francisco but even the further-afield Oakland were turning into bedroom communities for a wealthy tech class with fewer and fewer connections to the area's civic and (counter)cultural histories. By the mid-2010, this resulted in protests and, ultimately, an evidently unsolved series of pellet gun attacks on the buses.[11] This growing popular discontent was perhaps most distilled in a 2014 incident in which a woman sporting the experimental wearable computing platform Google Glass was assailed at a punk bar in San Francisco, briefly having the offending technology taken from her amid general expressions of anti-industry sentiment.[12]

At the same time, the increasing dominance of the region in everyday life everywhere that smartphones are available has broadened this discontent beyond local disputes and into the national and even global sphere. Social media companies have come under attack by both the right and liberal wings of American politics for their data collection and moderation policies (or perceived lack thereof). Abroad, and particularly in Europe, governments have increasingly sought to regulate their activities. As we have seen, reporting on the area has also taken an increasingly skeptical turn. The ubiquity of brands like Facebook, Google, Netflix, and Apple (and Twitter/X, based in San Francisco until 2024) in shaping not only technology but economics, politics, and discourse has seemingly effected a resurgence of critical standpoints on the Valley largely unseen since the early 1990s, and at a much broader scale.[13]

If the consolidation of data in high-end server farms has raised awareness of the concentrative tendencies of post-dot-com Silicon Valley, the global COVID-19 pandemic has exacerbated this situation by beginning to make good on the promise of a "placeless" society that the internet long heralded. No longer beholden to cubicles, some white-collar residents of Silicon Valley—and expensive states like California more broadly—are seeking more affordable pastures, raising rents and property values in smaller cities and magnifying declines in the regions they are leaving.[14] These trends, and the reluctance of employees to return to shared physical working locations, have led to some companies, even ones as insistent on in-person collaboration as Apple, to propose geographically diversifying their offices. Interestingly, according to one report, the company's silicon division has led the way on these initiatives.[15]

In this sense, the centrifugal force the region's labor market is currently subject to is an inevitable and dialectical reaction to the San Francisco Peninsula's centripetal consolidation of economic and political power over the past thirty years. Its influence, initially dependent on exponential growth in processing power and subsequently on the centralization of online services, may now be dispersing, at least spatially. As the material basis for which the Valley is named founders, the high costs associated with the area are beginning to appear unnecessary to workers and management alike. What began in a unique combination of expertise and geography[16] could end up as something closer to a state of mind, as both silicon and the Valley become a less critical part of an industry dependent more on user lock-in than technological advances.

Transistors, or, the Racialized Logic Circuits of Late Capitalism

As the renegotiation of labor conditions amid the increasing feasibility of remote work demonstrates, Silicon Valley has been at the forefront of many developments in not only the forces but also the relations of production. For Christophe Lécuyer, this is traceable at least as far back as Eitel-McCullough, Inc.'s early 1940s mass production of radar tubes for US military use in San Bruno, a city on the peninsula north of the Valley. In its newly built factory, which "increasingly hired women for delicate assembly operations," the company "fought vigorously against the unions," successfully mollifying its shop floor in "the largest and most active center of trade unionism in the Western United States" with "corporatist" methods, distributing profits to and discounting on-site meals for workers.[17] Similar efforts, Lécuyer writes, were made a few years later in what would become the Valley proper by microwave valve firm Litton Industries.[18] "In 1960, Fairchild Semiconductor," Moore's own company, "took the innovative step of giving stock options to some of its engineers" over the objections of its parent company, Fairchild Camera and Instrument, who feared "a socialist plot."[19] These approaches, of course, would go on to characterize high-tech enterprises in the region.[20]

Yet if this was socialism (it wasn't), it favored white-collar labor. As Lécuyer also notes, "the working conditions of technicians and operators could be grim."[21] When it came to factory work, Fairchild Semiconductor's management "viewed unionization as a death threat."[22] As early as 1958, the year after it was founded, the company had been advised by the president of Fairchild Camera "about choosing a site and size" for expansion "so as to avoid the prospect of a unionized workforce."[23] When Taylorism did not bring costs down enough, its leadership "vowed to shift assembly operations to areas where labor unions were weaker": first to Maine in 1962, then Hong Kong the following year, and South Korea a few years later.[24] Both in tandem with and separate from employee perks, anti-unionization was a key aspect of the industry from which Silicon Valley takes its name. It also went hand-in-hand with geographical diversification.

In an insightful essay, Lisa Nakamura traces the globalizing development of electronics manufacturing through a key moment: the "insourcing" of Fairchild assembly to Native American reservations—specifically the Navajo Nation territory of Shiprock, New Mexico—from 1965 through 1975.[25] This period roughly spans the development of Moore's law from its first publication to when Moore extended the doubling rate it described to the now-standard two years. It thus reflects a decade of particular dynamism with regard to not only the economic development of integrated circuitry but its cultural impact, one that, Nakamura argues, "provides a valuable perspective from which to view the material culture of computing."[26] While really drawn to Shiprock because of "tax incentives" and "the lack of unions and other employment options in the area," key for her is the way these impersonal economic advantages were glossed over in favor of the company's description of the workforce in its promotional materials. Racialized characterizations, she argues, were manufactured and leveraged to connect attributes like an ostensibly "inherent and inborn manual dexterity," supposedly already evident in Native American craft and especially the gendered practice of weaving, with the increasingly technical work of electronics assembly.[27] Such a move recalls the similarly "delicate" work of the female labor force at Eitel-McCullough.

For Nakamura, it also anticipates the neoliberal emphasis on flexibility we have already encountered in chapter 4. Even prior to the industry's efforts to globalize, she writes,

> immigrant women of color were hailed as the ideal workforce because they were mobile, cheap, and above all, *flexible*; they could be laid off at any time and could not move to look for alternative forms of work, while their employers could close plants and reopen them in locales with the most favorable conditions. The notion that Indians were "inherently flexible" both racializes and precedes the idea of flexible labor that informs much of the research on globalization in the information age.[28]

In this regard, it is especially noteworthy that the paradigm that developed out of Moore's law essentially requires expensive new plants to be constructed as manufacturing processes operate at finer and finer scales. The industry is thus also frequently presented with opportunities to reassess the "flexibility" of the labor it employs and discipline it if it is not sufficiently pliable. Relocating production, then, whether next door or halfway around the world, is a fact of business in terms of both fixed and variable capital.

The results of this continual reinvention, as with much of the globalization of production, have often been disastrous for the populations left behind. As Sean Cubitt argues, "Indigenous people have borne the brunt of the digital boom," and "the global poor suffer far more from pollution and environmental loss than the global rich," a situation playing out throughout much of the Global South as fabs continue to close in some places and open in others.[29] Nakamura's analysis concurs: "Semiconductor manufacture continued the ongoing practice of environmental degradation in a spot renowned for its natural beauty. Ultimately, the Navajo nation failed to benefit economically as much as it had expected from the plant and was left to deal with the detritus and its long-term consequences."[30] While the list of those subject to these precarious yet lasting conditions may have since expanded beyond these specifically racialized and gendered precursors, the case of Fairchild's Shiprock plant, she argues, nevertheless "reveal[s] potent and durable claims and beliefs about gender, race, and particular labor *styles* that would quickly be appropriated to describe the Asian women workers who eventually replaced them."[31] The contemporary distribution of labor within the industry, she concludes, still harbors echoes of this ideological move.

Xenofeminist critiques, likewise, look to "recent engagements with the digital that foreground its brute physicality over its supposedly more ethereal qualities" in order "to anchor that which has been frequently mischaracterized as free-floating and disembodied within its infrastructural requirements and within the obstinate physicality of its users and producers (including those workers engaged in repetitive and poorly paid labour on electronics assembly lines around the world)."[32] In this regard, it is important to disambiguate the different categories of labor within the semiconductor industry. Specifically, we must distinguish assembly from fabrication. This means differentiating, in Nakamura's terms, between Foxconn's mainland Chinese workforce, with its notorious labor conditions, and that of the Intel "bunny suits" responsible for creating silicon wafers and shepherding them into finished yields of chips, largely in Western countries or Taiwan.[33] The latter, while less highly remunerated (both proportionally, as we have seen, and in absolute terms) than chip design, remains a specialized and generally well-paying form of industrial labor compared to the former. If it is possible to argue that "on the spectrum of digital labor, factory work soldering chips for iPhones, missiles, and servers is as close to the machine as one can get, as close to the means of digital production—the computer—as can be imagined,"[34] it is only because the labor of semiconductor fabrication itself is much closer to what Marx famously saw as the logical endpoint of capitalist production in his "Fragment on Machines."

There, Marx describes employment as a condition where "in no way does the machine appear as the individual worker's means of labour . . . this activity, rather, is posited in such a way that it merely transmits the machine's work, the machine's action, on to the raw material—supervises it and guards against interruptions."[35] Indeed, amid the pandemic-related semiconductor shortage, such supervisory positions found themselves in high demand. As the *Wall Street Journal* made clear, "while chip makers have an advantage because their processes are among the most-automated, the high-tech equipment used at their facilities still requires skilled staff to operate it."[36] The comparatively highly exploited labor of assembly, or connecting parts together into a finished product, only takes place subsequent to the manufacture of the wafers themselves.

Put differently, as with artificial intelligence, this is not simply a matter of the human labor *behind* ostensible automation. It is also the ongoing *replacement* of certain types of labor by machinery, their removal from the living realm of valorization and entry into the undead demesne of fixed capital. Soldering, as labor, has moved from the internal components of what are now integrated circuits to the external connections between chips and even entire boards required to enable a fully functioning product.[37] Much of the functionality enabled by Shiprock workers has long migrated below the level of direct human interaction. The necessity for workers to wear special clothing to avoid even microscopic amounts of dust contaminating the ever-shrinking fabrication process testifies to this. In an even more extreme manner than we have seen Kittler argue about their design since 1971, processor manufacture is now almost entirely mediated via computerized automation. Indeed, one xenofeminist response to the ongoing existence of "female workers in electronics . . . perform[ing] some of the worst paid, monotonous and debilitating labour" is "structural, *machinic* and ideological correction."[38] Extending the automation already observable in semiconductor fabs to the subsequent assembly tasks still completed by gendered and—as we have seen—racialized, human labor, is thus, for them, on par with the redistribution and reconceptualization of this labor itself.

This distinction between types of labor is also significant for another reason: the further racialized exclusion of certain workers from even this lower-tiered category—indeed, from labor as such. This is the very situation those temporary beneficiaries of globalization's restless movement set out to avoid by welcoming its productive forces to their shores. Here, too, Silicon Valley has been a pioneer, this time not through the work of Fairchild but rather the head of the company Moore and others left to found it: William Shockley. Shockley is famous within the electrical engineering community for the Nobel-winning transistor research that he and two of his supervisees at Bell Labs carried out in the late 1940s. He was also primarily responsible for the establishment of Silicon Valley, founding, together with Beckman Instruments, Inc., the Shockley Semiconductor Laboratory in the mid-1950s. It was this firm that Moore and Robert Noyce, who together would ultimately go on to start Intel, first left (with six others) to establish Fairchild.

Shockley is equally infamous, however, for his research into and vocal support for eugenics pursued, after leaving his business ventures, as a faculty member at Stanford University. Nakamura's analysis of Fairchild in fact began as a project to investigate Shockley's racist ideology,[39] though these two entities' attitudes toward race should not be conflated. Indeed, as she writes, her research "started out as a project on eugenics and Silicon Valley and ended up as something very different."[40] Shockley's racism, which became increasingly explicit over the course of his life, has, until recently, often been glossed over in the technical literature. For example, in an overview of his "inventive genius" in the same issue of the *Proceedings of the IEEE* in which he reflects on Moore's law, Probir K. Bondyopadhyay elides most of Shockley's efforts outside of engineering except to describe the formative influence of his observation of the famine in Bengal during the Second World War. He writes, "Highly sensitive to humanitarian concerns, Shockley's thoughts and activities during the last third of his life were triggered and shaped by this wartime experience," but, beyond a citation to one of Shockley's treatises on eugenics, does not dwell on what those "thoughts and activities" were.[41]

The conceptual firewall between Shockley's engineering accomplishments and his preoccupation with race and genetics was undoubtedly made easier to maintain by the latter's full-throated emergence following his return to academia. Shockley's retreat from industry occurred just before the region transformed itself into a center for not only electronics but computation proper. He may have helped invent the transistor, in other words, but he arguably had less to do with the computer as Silicon Valley has subsequently led us to understand it. A closer examination of Shockley's thinking reveals this separation to be not only a matter of historical whitewashing but one that has served to erase revealing linkages between these two aspects of his career.

To begin with, it is important to situate this intellectual trajectory within a larger climate of industrial automation. As David C. Brock has shown, both Shockley and his business partner, Arnold Beckman—a fellow Bell Labs alum whose company had become a pioneer in chemical analysis equipment, expanded into analog computers and control systems, and aimed to diversify further—were immersed in the heady reconfiguration of postwar production around cybernetics and semi-autonomous control systems.[42] This included a mid-1950s contribution from Shockley to the development of components for the Pasadena-based Consolidated Electrodynamics Corporation's new digital computer.[43] While we have already seen Norbert Wiener introduce the concept of the "automatic factory," Beckman and Shockley's attempts to extend it, Brock argues, were the cornerstone of their relationship and thus essential to their foundational, but ultimately unsuccessful, Mountain View enterprise.[44] At the same time, there is some evidence that the two men approached this topic from slightly different perspectives. Beckman at least paid lip service to the notion that automation would be a general good, writing in his company's inaugural 1952 missive to those who held stock that "'machines liberate men.'"[45] Shockley, on the other hand, pushed further into the territory of outright redundancy, proposing a Bell Labs project around "'trainable robots'" for "the substitution of machines for men in production."[46] While positioning them as the answer to perceived labor shortages, including in the field of housework, rather than a direct driver of unemployment, as Brock notes, "organized labor and cultural critics had voiced concern about . . . automation. For them, Shockley's version would be far more threatening."[47] Well before Fairchild's efforts at outsourcing, then, automation and its effects on labor occupied a significant place in the nascent semiconductor industry's imaginary.

Paradoxically, Shockley, having negotiated personal ownership over one of his patents in this area, saw the development of unemployment-inducing human substitutes as the key to his own escape from waged exploitation and entrée into the ownership class.[48] His proposition, however, was ultimately rejected, contributing to his decision to form his own company.[49] "He wanted a new life," Brock writes, "and a company of his own through which *he* would be the primary beneficiary of his labors," this ambition too becoming a foundational aspect of Valley culture.[50] Eventually falling in with the like-minded Beckman, Shockley's resulting venture, however, would seemingly not emphasize these topics at all. Though automation was key to the commoditization of silicon transistors they aimed for, in his draft heralding the establishment of his company, Brock notes, "nowhere did [Shockley] mention any need for or interest in hands experienced in high volume, automated, or semiconductor manufacturing."[51] Yet, far from representing distinct areas of concern, his attempts to "scientifically" establish the superiority and inferiority of various genetic "ethnic" groupings were informed by this context. As we will see shortly, they initially arose from a concern with *labor*. Crucially, however, Shockley's eugenic approach to both this topic and, subsequently, race was informed by the earlier work that he and the others at Bell Labs had done to bring the semiconductor revolution into being.

For Shockley, the development of the semiconductor and the socioeconomic impact of new technologies like it legitimated a novel approach. In contradistinction to classical understandings of racism as focused on genetic "hygiene," questions of *im*purity were key to both Shockley's groundbreaking work in semiconductors *and* his thinking on eugenics. The work for which Shockley's crew won the Nobel Prize in 1956 was made possible by a careful mix of purity and impurity: exactingly undiluted germanium "doped" with a different agent—in this case, tin—to produce an *N*-type, or negatively charged, semiconductor.[52] Shockley's own contribution was the junction transistor, a layered approach situating *N*-type semiconductivity between positive, or *P*-type, areas, adding even further impurity and hybridity.[53] With this method, it was a very specific *admixture* of elements that resulted in the unique properties for which semiconductors came to be known, their specific conditioning, or rectification, of electrical current. Progress was achieved not by relentless refinement of an exceptional element, but rather the right *combination* of elements: a very postmodern concept.

When Shockley's attention turned to genetics and race, these same presuppositions held, despite the radically different domain. In the abstract to a 1966 paper presented at the National Academy of Sciences published in *Science*—one of his earliest assays into the area of race after he had made the transition from industry to academia—Shockley posits the applicability of the conceptual breakthroughs in chemistry that enabled the transistor to the nature vs. nurture debate:

> The analogy proposed here makes impure "chemically pure" elemental metals correspond to impure "genetically pure" populations and makes a radioactive tracer atom correspond to a tracer gene G_i—for example, blood type or taste discrimination—that occurs in different fractions a_i and b_i of individuals in "elemental populations" P_A and P_B. A large group of G's can be used to measure, for each individual studied (rather than for populations), the "ethnic composition fractions" f_A and f_B of the individual's genes that come from P_A and P_B ancestors.[54]

The method he describes has parallels in not only the design but the subsequent production of semiconductors. Radiotracer-based diffusion studies, for example, had been significant to ensuring the integrity of transistors' hybridity.[55] Drawing on these applications, Shockley was positing that collections of individual genetic traits could be summed up, across a broad enough range of samples, into cohesive ethnicities in much the same way a combination of chemical elements could be designed and verified as semiconductive.

If this approach were valid, he wrote, it would mean that whether "the expectation value of an individual's performance is independent of ethnic composition (that is, ethnic heredity is trivial compared with environment) should be possible to establish or to reject to a high level of statistical significance by observations on siblings of mixed P_A—P_B ancestry."[56] While Shockley was being circumspect here, his own leanings, as we shall see, ultimately favored the heredity argument—that is, the notion that some "ethnic" groups were more highly "performant" than others. This conclusion would situate Shockley as an heir to previous—and ultimately false—efforts at "scientific" racism such as phrenology and early twentieth-century eugenics. Yet his earliest exploration of these concepts—and thus the pivot point between his previous work and subsequent eugenics advocacy—was not in the direction of race but rather toward the problematic of labor, specifically the role of labor in an emerging so-called knowledge economy.

In a 1957 paper—the same year the "traitorous eight," including future Intel cofounders Moore and Noyce, would leave Shockley Semiconductor to start Fairchild—Shockley attempted to account for disparities in research productivity at large Cold War–era labs through this lens. He did so by attempting to quantify individual researchers' contributions to publications and patents. Immediately notable here is his effort to demonstrate that across a given laboratory's research force a "log-normal" distribution broadly holds; that is, when charting individual researchers' productivity on a logarithmic scale, a roughly linear relation is represented, with most researchers situated between extremes of low and high productivity.[57] While the visual similarities to Moore's famous logarithmic graph are plain (see figure 5.1), their implications are almost exactly opposite. Shockley's thesis was synchronic where Moore's law is diachronic. Where the latter indexed engineering improvements to the passage of time, as straightforward a definition of progress as one could hope for, the former represented a fixed order with the most productive on top. The very tool of exponentiality itself can thus be seen as a contested method in the establishment of Silicon Valley ideology. It could be marshalled to justify the inevitability of either technological change or indelible hierarchy, the utopian or dystopian strand of the future the region was building.

Shockley's justification for this distribution—and thus the basis of the social order he favored—was, as with Fairchild's reasoning for opening its plant in Shiprock, an appeal to "nature." However, Shockley thoroughly infused this concept with the logic of hybridity informing his breakthrough technical discoveries. Like the logic of neuronal intelligence, it hinges on connections between varying elements, though crucially here these are not material factors of synaptic mediation but rather conceptual relations. "Let us suppose," he wrote, "that there is some attribute of the human brain which allows an individual to be aware of '*m*' ideas and their relationships. Then it follows that a man with m=3 will never invent [a device combining four ideas] whereas a man with m=4 can do so."[58] What could account for this *m*, he posits, is "the hypothesis that the interacting mental factors are of several different kinds rather than several of one kind."[59] Indeed, the cumulative interaction of multiple minute changes, as in the previous development of the semiconductor and his subsequent genetics abstract, can produce perceptible results: "If we suppose that these factors vary independently, then to a good approximation their sum will have a normal distribution and so, consequently, will the logarithm of the

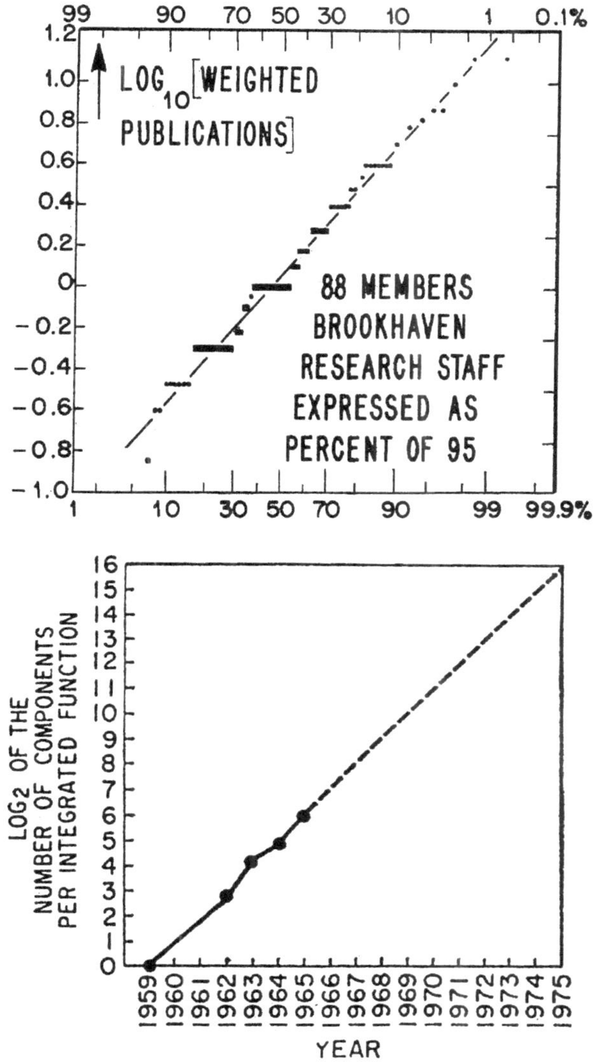

Figure 5.1

Two opposing uses of the semi-log plot: William Shockley, "On the Statistics of Individual Variations of Productivity in Research Laboratories," *Proceedings of the IRE* 45, no. 3 (March 1957): 283 (top) © 1957 by the Institute of Radio Engineers, Inc. Reprinted with permission from *Proceedngs of the IRE*; Gordon E. Moore, "Cramming More Components onto Integrated Circuits" (1965), reprinted in *Proceedings of the IEEE* 86, no. 1 (January 1998): 84 (bottom) © 1998 by the Institute of Electrical and Electronics Engineers, Inc. Reprinted with permission from *Proceedings of the IEEE*.

productivity."[60] For Shockley, exceptional researchers, like transistors, are made from rare amalgamations of independent elements.

The relationship between labor and race here becomes clear when we consider the focus across these two papers on performance and productivity, particularly in the intellectual sphere. The 1957 paper, of course, explicitly attempted to measure intellectual output, perhaps informed by Shockley's earlier sense that factory labor could already be accounted for by machinery. However, Shockley's 1966 abstract also culminates in a suggestion that "correlation of physical, mental, and social measurements with one another and with values of [his 'ethnic composition index'] *H* might give scientific support for [Berkeley anthropologist Sherwood] Washburn's proposal (that is, for equal environments American Negroes [*sic*] might surpass whites), by showing a positive correlation of performance with fraction of Negro [*sic*] genes."[61] Washburn's suggestion, however, specifically related not to physical or social metrics but comparative performance on IQ tests (which, of course, have their own well-documented history of eugenicist influence and racial bias).[62] Shockley thus ended his abstract by proposing that "Data on correlations of *G*'s would be genetically significant and relevant to I.Q. and other polygenic traits."[63] What had previously presented itself in Shockley's research as a matter of scientific labor productivity was now turned toward racial discrepancies.

An interview with Shockley in *U.S. News & World Report* the previous year—which generated opprobrium even at the time—reinforces these preoccupations. There, he called for more study of potential genetic factors in the gap between average white and Black IQ scores: "We lack proper scientific investigations, possibly because nobody wants to raise the question for fear of being called a racist."[64] Moreover, the coinventor of the transistor justified the necessity of taking genetics into account partly on the basis of an increasingly complicated and technical mode of production. Asked bluntly by his interviewer whether "a society becoming more and more technological [can] afford to continue having large numbers of defective and dull people in its population," Shockley responded, "Certainly not. There will be less and less work that such people can do, and less and less that they will be able to comprehend in the world around them."[65] In this worldview, unionization was less of an issue than the way that hierarchies of labor and race were intertwined around the measure of intelligence. On the one hand, value creation is recognized as demanding an increasingly large cognitive component, leading Shockley to echo his former partner's call for better STEM preparation in the same venue nearly a decade earlier.[66] On the other, genetic traits, conceived as legitimators of racial hierarchy, are posited as determinants of access to this education[67] and success within these progressively important job categories.

In a sense, Shockley's worries of being labeled racist were valid, because he would demonstrate without a doubt that he *was*. What ostensibly appeared as plausible deniability and intellectual "courageousness" from a Nobel winner in the mid-1960s foreshadowed unabashedly racist conclusions, affiliations, and statements made over the rest of his life.[68] The so-called humanitarianism that drew him to entertain Malthusian solutions to hunger in his wartime years eventually aligned him with explicit white supremacists, leading to a failed 1982 Senate run intending to "use his campaign to explain his view that blacks and some other races are not evolving as quickly as others," and ultimately, until his death in 1989, a kind of privileged exile as a largely shunned emeritus professor.[69] There is much more to say about Shockley's ideas, both technological and racial, than there is room to do so here. For the present purposes, however, it is highly significant that both the material basis for the advent of the "knowledge economy" as well as the ideological justification for it, predicated upon first labor- and subsequently race-based conceptions of mental capability, should stem from not only the same location but the same person.

In other words, the seeds of a postmodern, globalized, and racialized division of technological labor had already been planted *before* Fairchild began region shopping for lower-paid and more pliable workers. Moreover, they had been sown by the very person with whom its founders had parted ways in order to do so. This division was not fundamentally predicated on Native American or Asian manual labor. It was instead founded on highly dubious methods of assessing intelligence that authorized a perception of Black Americans as *unsuitable* for both these *and* more white-collar positions within such industries. Shockley's paper on research productivity was published near the all-time high of US workforce unionization.[70] His conception of high-tech labor sought validation nine years later, just as older structures of racial exclusion—most notably explicit segregation—were beginning to crumble. Shockley's understanding of hybridity therefore both ushered in the shift from industrial to postindustrial technology and legitimated its brutal "meritocracy" based on intellectual proficiency.[71] This is by no means the only influence on the Valley, which, following Beckman, often touts and sometimes even seems to genuinely believe that it is in the business of improving things for everyone.[72] Nevertheless, appraisals of cognitive capacity still play a role in its 2020s culture via the use of services like LeetCode to assess programming proficiency, while manual labor remains a distant concern. In solidifying this concept, Shockley's industry heirs may have rejected his most unsavory ideas, but his emphasis on the assessment and categorization of worthiness remains central.[73] While his explicitly racist writings have unsurprisingly been disavowed, Shockley's research performance study is still cited today.

The emphasis on statistical rates of intelligence across a population and within racialized subgroups is, of course, evidence of what Michel Foucault has called "'biopolitics' . . . the attempt, starting from the eighteenth century, to rationalize the problems posed to governmental practice by phenomena characteristic of a set of living beings forming a population: health, hygiene, birthrate, life expectancy, race," such that "security mechanisms have to be installed around the random element inherent in a population of living beings so as to optimize a state of life."[74] Biopolitical government, Foucault argues, must practice a type of racism of which eugenics is exemplary: "The death of the other, the death of the bad race, of the inferior race (or the degenerate, or the abnormal) is something that will make life in general healthier: healthier and purer."[75] However, Shockley's work discussed here offers a specifically postmodern racism, in that "purity" is defined as optimal hybridity. It also offers a postindustrial form of biopolitics, one that seeks optimization of a shrinking labor force and thus regulation of "surplus" populations.

From a Marxist perspective, then, Shockley's racism is noteworthy not only due to the way he attempted to justify it via technological discovery. It is significant because his approach to race stems out of his theorization of labor and its implications for the labor *market*. These developments, in turn, are influenced by an increasingly automated post-industrial economy enabled, at least in part, by that very same technical innovation, the transistor. As Craig Wilse has argued,

> looking back we can see that [Foucault's] model of biopower charts the emergence of the modern nation-state, especially in its early twentieth-century welfare state form. In the context of a Keynesian political economy, the life of the national population is set to positively correlate with the life of the economy: both must

> grow for either one to grow. In a post-Fordist, deregulated capitalism, the economy becomes divorced from labor; the abandonment of mass segments of the US population (through the dismantling of social welfare programs, increased privatization of healthcare, defunding of public education) is not only social, but, very literally, economic. The economy moves on without many of us.[76]

A theory of 100% utilization of course holds, as a Marxist analysis cannot but do, that it is ultimately impossible to dissociate economics from labor. Leaving this momentarily aside, however, there is no contradiction with the rest of Wilse's statement and the analyses of lower employment rates offered here. As Foucault observes, shifts in unemployment policy to focus on temporary joblessness for any reason, rather than more sustained job loss due to specific circumstances (like the shuttering of a factory), characterize the onset of the neoliberal era.[77] The concern, for both Shockley and the state, is less how to prevent worker organization than which of the masses will be fortunate enough to get a stable job in the first place. A postindustrial state biopolitics must manage a growing reserve army, one it may never need to call up thanks at least in part to the proliferation of semiconductor-based methods of (re)production, simulation, and distribution alongside a shrinking welfare state. In this sense Shockley's approach to race and labor—and race *through* labor—both reflects how, as Foucault notes, "the specificity of modern racism . . . is bound up with the technique of power, with the technology of power,"[78] and indexes the changes in this technical foundation.

Dialecticizing the Singularity

That such technological transformations have subsequently overflowed the boundaries of race-based categorization even as they have left many racial inequities intact ought not be surprising. Photographer Mary Beth Meehan and Stanford scholar of tech culture Fred Turner illustrate this dual development in their portrait of inhabitants from a range of class and identity positions struggling with the pressures of life in the post–Shockley Valley. As they report, in a local town hall one Black townswoman challenged Facebook head Mark Zuckerberg, "'What about the young people, those who need jobs, from *within* this community?'"[79] As with the transposition of techniques of colonial domination from what was once the periphery to the former center,[80] or, perhaps more directly, the increasing application of neoliberal austerity measures once reserved for those same now post-colonies to the developed Western nations that had imposed them, people across racial boundaries are increasingly confronted with the exclusionary logic of a "knowledge economy" seeking to transfer more tasks onto machinery even as it trumpets efforts to diversify what remains of its workforce in ways that privilege a transnational elite over preexisting, historically marginalized groups.

This "meritocratic" legacy of Shockley's ideas is, as Ruha Benjamin writes of modern-day eugenic efforts to inform prospective parents of all races about the DNA makeup of their embryos, "yet another example of how those who design the world according to their own values and biases are employing the rhetoric of 'inclusion' as progressive veneer for deeply discriminatory practices."[81] Wendy Hui Kyong Chun has argued that this type of so-called positive eugenics—that is, eugenics aiming at promoting specific kinds of life rather than extinguishing those it deems undesirable—is a model for a conception of computation by which material operations convey instructions for optimal execution. "Breeding," she writes, "encapsulates an early logic of programmability that inspired genetics and recognition of heredity as physical transmission. Eugenics, in other words, was not simply a factor driving the development of high-speed mass calculation at the level of content (the statistical demands of the biometricians helped foster mass calculation), but also at the level of logic or of operationality."[82] Yet Shockley's bridging of electrical engineering and explicitly racist eugenics makes clear that programmability is not strictly necessary for this connection to be made. It can already be identified in the concrete makeup of the very first transistors, components that form the fundamental basis of "high-speed mass calculation" but which were only subsequently understood as programmable. As their performance at this task has improved, however, more and more people across racialized categorizations have fallen under the logic of exclusion amplified by the trade-off between computation and labor power they enable.

The ultimate fantasy underlying this tendency arrives, paradoxically, when the expulsion that Shockley sought to legitimate becomes total and the technologies for replacing those it has shut it out reach impossible heights. This is the "technological singularity": the hypothesized moment when computers become autonomously intelligent, surpass their creators, and usher in a world where the latter no longer need to work, for either good or ill. Indeed, as Brock reminds us, the "enthusiastic vision" and "tremendous consequences" found in both the positive and negative prognostications associated with the 1950s "automation movement" are "similar to other sociocultural movements . . . that formed around scientific and technological developments: information theory, strong artificial intelligence, recombinant DNA technology, chaos theory, molecular nanotechnology, personalized medicine, and synthetic biology,"[83] all of which are key touchstones for devotees of the singularity. While the concept itself did not originate in the Valley, the region and the Bay Area more broadly has both produced its own proponents and concentrated those from outside. Sun Microsystems cofounder Bill Joy, futurist entrepreneur Ray Kurzweil (now employed by Google), and superintelligence soothsayer Eliezer Yudkowsky are some of the most well-known voices in this arena. They have also built connections within academia. Yudkowsky's Machine Intelligence Research Institute, for instance, launched the inaugural "Singularity Summit" in 2006, including among its speakers Oxford reality-as-computer-simulation theorist Nick Bostrom, whose immaterial vision of an infinite cascade of virtual worlds, as we saw in the introduction, runs afoul of the truth of computation as a concrete procedure.

Similar miscalculations dominate singularity discourse, an outcome as unlikely as the simulation hypothesis and for much the same reason. Joy's predictions for this scenario have been the most apocalyptic. In an influential 2000 article published in *WIRED*, he argued that technological change has inured us to the consequences that developments on the horizon might hold. Contending that "because of the recent rapid and radical progress in molecular electronics—where individual atoms and molecules replace lithographically drawn transistors—and related nanoscale technologies, we should be able to meet or exceed the Moore's law rate of progress for another 30 years,"[84] he famously offered a warning about this degree of exponential technological improvement. Specifically, Joy projected a world taken over by mechanically self-reproducing nanomachines—a technology probable, in his then-estimation, by the year 2020—that could ultimately lead to their dominating the biome: the so-called "gray goo problem."[85] To the extent that it emphasizes similar hazards with genetic engineering, Joy's essay remains relevant for a world emerging from a pandemic of uncertain origin. He even warns of the dangers presented by global capitalism and an ideology of endless growth. Nevertheless, to the degree that he overestimates the extent of computational development, does not anticipate the issues with Moore's law that have since become clear, and assumes that curtailing growth must be a social choice rather than an external limit, he participates in the kind of technological boosterism Brock describes, as pessimistic as it may be.

Joy's piece concludes with a call for a more even-handed assessment, but this same tendency is in even greater evidence for many comparatively nuanced—indeed, more dialectical—approaches. This includes those championed by the Singularity Summit, which sought to consider both the promise and peril of an unbroken line of exponential computational progress. The now-archived website for the initial conference, held at Stanford nearly seventeen years after Shockley's death, invokes a twentieth-century Malthusianism that would not have been unfamiliar to him. It also illustrates the summit's linkages with the Bay Area Technorati via another connection with *WIRED*, a quote from its "Senior Maverick," Kevin Kelly:

> With his new book, *The Singularity Is Near*, Ray Kurzweil has taken all the strands of the singularity meme circulating in the last decades and has united them into a single tome which he has nailed on our front door. I suspect this will be one of the most cited books of the decade. Like Paul Ehrlich's upsetting 1972 book *Population Bomb*, fan or foe, it's the wave at the epicenter you have to start with.[86]

From our perspective, however, *The Singularity Is Near*, like *The Population Bomb* and unlike the Lutheran Reformation, appears to have overstated its case. Kurzweil's book, one of his many statements on the ever-incipient arrival of superintelligence, is predicated on always-advancing computational capacity in contradistinction to what he sees as the fixed, or at least slow-to-change, capabilities of the human mind. Moore's law figures intriguingly into this argument as evidence of the former celerity that he nevertheless refuses to let bind his thesis. "When Moore's Law reaches the end of its S-curve," he writes, "now expected before 2020, the exponential growth will continue with three-dimensional molecular computing, which will constitute the sixth paradigm."[87] The good news for this prognostication is that Moore's law has not yet reached this terminus. The bad news is that this is in large part due to engineering difficulties keeping it behind its strict exponential schedule. Moreover, as we have already seen, Kurzweil's "sixth paradigm," after computational epochs predicated upon electromechanics, relay switches, electron tubes, discrete semiconductors, and integrated circuits, respectively,[88] is not yet a practicality, despite moves to shift integrated circuit architecture into three dimensions.

This disavowed reliance upon Moore's law allows Kurzweil to evade simplistic dismissals, but its ideological character is revealed by proper historicization. In his 1990 book *The Age of Intelligent Machines*, he outlines similar possible futures for the material basis of computing as he does in *The Singularity Is Near*, but also emphasizes that "a hardware breakthrough . . . is not *necessary* for significant and sustained progress in each area of AI . . . the exponential growth in computer power that will occur anyway is sufficient."[89] By 2013, while still insisting that "three dimensional computing will keep that exponential trend going for many more decades," he omits the "molecular" aspect—the reduction in size that most terrified Joy—of his 2005 prediction.[90] Despite Kurzweil's relentless belief in the ever-growing power of computation, the state of Moore's law at the moment each of these pronouncements is made, from the wide-open future of 1990 through the beginning of clock speed stagnation in 2005 (even as he deploys historical data to cite a clock doubling rate of three years)[91] to the beginning of the troubled 14 nm era, would seem to have a palpable effect on their contents.

The Valley and the class interests it protects, however, do not hold a monopoly on this kind of flawed hopefulness. It has increasingly become a significant part of a new-generation left-wing discourse as well. Drawing in many ways from the legacy of 1990s thinkers affiliated with the British Cybernetic Culture Research Unit, it, too, offers a perplexing sense of constancy where one cannot shake the suspicion that a more historical analysis should obtain. Thinkers affiliated with the CCRU, confronted with the untrammeled global expansion of capitalism and accompanying breakthroughs in popular computing following the 1989 collapse of state socialism, embraced the contradictions and possibilities this post–Cold War environment offered. For example, they championed the accelerated drum samples and digital wizardry constituting the some of the most technologically advanced music not only of the time but to date: jungle/drum'n'bass.[92] For its inheritors, the presuppositions that legitimately made this previous moment so heady—that global culture was careening toward a new, ostensibly posthistorical future based on remixing and other forms of electronically enabled intercommunication—have been largely extended to describe our more recent conjuncture, with its subsequent yet vastly different historical context, under the heading of "accelerationism."

According to former CCRU member Robin Mackay and Armen Avanessian in the introduction to their collected volume exploring the tradition, "At the basis of all accelerationist thought lies the assertion that the crimes, contradictions and absurdities of capitalism have to be countered with a politically and theoretically progressive attitude towards its constituent elements."[93] This may seem like a perfectly dialectical viewpoint, and one also found in Marx—indeed, his "Fragment on Machines" is included in the collection. However, in practice this approach often proceeds from the premise that technological progress continues in some sense independently from the material and social conditions under which it appears. This, in turn, makes technology appear a limitless reserve of societal change.

As accelerationists Alex Williams and Nick Srnicek argue, "Our wager is that the true transformative potentials of much of our technological and scientific research remain unexploited, filled with presently redundant features (or preadaptations) that, following a shift beyond the short-sighted capitalist socius, can become decisive."[94] Related manifestos go even further when they argue that Moore's law's impending end is simply a speed bump on a trajectory of ever-improving computational power. Aaron Bastani's *Fully Automated Luxury Communism*, which argues that nascent improvements in technologies like AI will usher in an age of both abundance and unemployment, heightening the urgency of political transformation, offers an optimism on par with Kurzweil's. "While within a decade it may become impossible to miniaturise individual transistors any further, simply because of physical limits," Bastani writes, "adaptations such as 3-D circuitry and quantum computing—both proven concepts—could mean exponential growth continues."[95] Whether being held back or ushered in by capitalism, revolutionary technological change is always on the horizon for the accelerationists.

Yet this is only one way to read Marx's famous description of "social revolution," that "at a certain stage of development, the material productive forces of society come into conflict with the existing relations of production or—this merely expresses the same thing in legal terms—with the property relations within the framework of which they have operated hitherto."[96] There is, after all, a long trajectory of thought, stemming from the Frankfurt School, which holds that capitalist political-economic development reached its peak in the early twentieth century, with the failure of a Weimar-era socialist revolution marking the onset of its self-cannibalization.[97] In such a theorization, subsequent improvements in fixed capital, at least beyond any postwar reconstruction, would be comparatively insignificant. Indeed, this is roughly Aaron Benanav's thesis, from whom we shall hear more imminently. Yet the period following the end of the Second World War also ushered in a new type of machine: the discrete, electronic computer. It is therefore possible that the stage for another, potentially more dispositive, clash has been set by the consequent shifts in the mode of production explored throughout this book—that is, by a digitized economy's lower need for labor and the heightened ease of machinic replacement it has enabled.

As creative capacity—that is, human labor—has been further diminished by digital capitalism it is not necessarily the case that we are awaiting further technological breakthroughs to radically intensify this dynamic. Instead, we could simply be treading water as the historical trajectory of technological progress itself decelerates. This possibility is what distinguishes a theory of 100% utilization from the accelerationist analyses we have encountered here and in Srnicek's analyses of platforms in chapter 2. It would explain the persistent sense of overall cultural stagnation diagnosed by those in the more explicitly Marxist ex-CCRU faction[98] as well as its more specific manifestation within Silicon Valley and its culture. It would also locate the material basis for those superstructural conditions, at least in part, in the slowdown to Moore's law.

The material condition infinitely more familiar in the Marxist tradition than the slowing rate of computational advancement is, of course, the declining rate of profit.[99] It is now the moment to reconcile—lest a theory of 100% utilization fall into the same trap of technological determinism haunting the champions of computational acceleration in the Valley and postcontinental philosophy alike—these two seemingly distinct factors. For there is, lastly, an opposite tendency to these exponential fantasies worth examining, not to endorse but to proffer as their dialectical inverse. It is represented most forcefully in Benanav's *Automation and the Future of Work*, which contends that crises of overproduction, rather than automation and technological advancement, are the major drivers of stagnation and un(der) employment. Essentially, Benanav argues that there *already* exists a surfeit of productive capacity *within* the world market—think factories—that has driven down the cost of goods. In this he concurs with Srnicek (both draw on Robert Brenner for their analyses), but the knock-on effects he traces lead him to a substantially different conclusion:

> The tendency to economy-wide stagnation associated with the decline in manufacturing dynamism, then explains the system-wide decline in the demand for labour, and so also the problems that the automation theorists cite: stagnant real wages, falling labor shares of income, and so on . . . the mistake of the automation theorists is to assume that productivity is rising at a rapid pace; whereas in fact, output growth rates have declined sharply over time.[100]

Such oversupply results, indeed, in a declining rate of return, dampening investors' enthusiasm in developing plant, which then affects employment and wages. This generates effects on the demand side: the inability of un(der)employed workers to sustain high enough consumption to put strain on factories prevents the kind of automation that would, in Benanav's view, actually raise commodity production trendlines.

Such an argument reinjects a desperately necessary acknowledgment of the stagnation and decline of recent decades that is seemingly as obvious to workers as it is invisible to those obsessed with the dwindling size of the pinouts upon which a superintelligent AI might dance. It is formulated explicitly as a rebuttal to what Benanav calls "the automation discourse" propounded by groups including the left-accelerationists, more mainstream liberal handwringers over the downsides of technological advancement, and "the jeans-wearing elite of Silicon Valley."[101] In (rightly) rejecting this techno-utopianism, however, his own approach to digitization is riven with contradictions, particularly around the question of computing power. At times Benanav is skeptical, listing doubts about the efficacy of AI and noting, as an aside to a brief consideration of the singularitarians, that Moore's law may be ending.[102] These positions presumably influence his argument that econometrics overvalue computing power. "Official US manufacturing growth-rate statistics," he writes, "are vastly overinflated, since they log the production of computers with higher processing speeds as equivalent to the production of more computers."[103] Yet computation is, at the same time, integral to his own vision of a postcapitalist society: "Some tasks would need to be performed locally, but many could be planned on a regional or global scale, using advanced computer technologies."[104] Benanav thus takes recourse to the least complicated answer to the productivity paradox, namely that digital technologies simply have little effect on economic growth—so much so, in fact, that they are even responsible for distorting this fact due to the way they are counted. In attempting to resolve the contradiction in this way, however, he runs afoul of another one.

This is, of course, the notion that computing power is simultaneously unimportant to optimizing our current mode of production, which already makes heavy use of it, yet key in some unspecified way to doing so in one that has not yet arrived. Such an antinomy, which goes far beyond Marx's exhortation that "the field of application for machinery would . . . be entirely different in a communist society from what it is in bourgeois society,"[105] represents a limit point in this mode of analysis. As a way of untying this conceptual knot, we can begin by taking seriously what Benanav dismisses: the economic impact of computing power. As we have seen, the 2011 National Research Council report on computation I have drawn on throughout this book warns of great industrial disruption if the growth trends established by Moore's law falter. Amdahl's law, meanwhile, examined in the previous chapter, provides a computer-scientific basis for the notion that processor speeds are ultimately more important than the raw number of commodity systems produced. It demonstrates how ultimately nothing but more speed can make up for single-threaded bottlenecks in even the most massively parallelized algorithms. The computer manufacturing figures that Benanav criticizes are therefore neither inaccurate nor misleading. Computing power really is an economic force in and of itself, and this fact troubles strict "productionist" accounts of political-economic development based primarily on counting the number of commodities manufactured.

We have seen how the ideological acceptance of past technological performance as indicative of future results leads accelerationists to entertain fantastical scenarios both dystopian and utopian. Benanav's orthodox insistence on the primacy of physical commodity production makes the opposite error. It overlooks the contributions that computation has made to staving off crises in the manufacturing sector, whether by reducing competition through consolidation, facilitating financing, improving industrial production (even if at a lower rate than previous innovations), or replacing material objects with electronic approximations. The extent to which digital technologies can achieve any of these ends is dependent on their capabilities, and thus computing power in general. To adequately grapple with the aspects of capitalist economy that have been informationalized, processing speed needs to be understood as quantitative. An analysis focused solely on industrial production, however, can only see it as qualitative, one feature of a certain number of objects made. Because of this, it also disregards the potential consequences for physical production—and indeed the economy as a whole—when Moore's law runs out. Seen from this perspective, the statement that "some commentators describe the present period of economic stagnation in terms of an exhaustion of the frontiers of technological advance—as if there were nothing left to invent—but it is more likely that low rates of industrial productivity growth are the result of a slower pace of industrial expansion rather than the reverse"[106] paradoxically aligns Benanav with those, like Williams and Srnicek, to whom he opposes himself. Both are ultimately dependent upon the notion that a surplus of technological innovation remains intact.

The theoretical basis for the tendency toward a falling rate of profit in capitalist societies, as described by Marx, is "just another expression for the progressive development of the social productivity of labour which is shown by the way that the growing use of machinery and fixed capital generally enables more raw and ancillary materials to be transformed into products in the same time by the same number of workers, i.e. with less labour."[107] In practice, however, as Marx subsequently notes, the vicissitudes of this tendency are more complex.[108] To begin with, it does not straightforwardly describe an increase in unemployment. Rather, it follows a decline in the ratio between variable and constant capital, human labour and the technology and inputs with which it works. This reduction can occur "even while both portions grow in absolute terms."[109] Indeed, "This not only *can* but *must* be the case—discounting transient fluctuations—on the basis of capitalist production."[110] However, there are two notable considerations here. Firstly, such growth is also compatible with a growing unemployed reserve army such "that the increased mass of means of production designed to be turned into capital finds a correspondingly increased and even excessive working population available for exploitation."[111] That is, even an *absolute* gain in the number of employed, when combined with a *larger* one in constant capital investment, can result in a relative decrease in the *ratio* between those with work and those without. It is worth noting, in this regard, that digital "dematerialization" would represent a qualitative leap in "the disproportion between the progressive growth of capital and the relative decline in its need for a growing population," since it removes manufacturing (and distribution) jobs almost entirely for some types of commodities.[112]

Secondly, "transient fluctuations" in unemployment do in fact play a substantial role in the tendency of the rate of profit to fall. In fact, Marx outlines what is essentially a macroeconomic circuit representing the course by which this happens. The model here is a kind of 100% utilization of capital, or "absolute overproduction of capital," where "no further additional capital could be employed for the purpose of capitalist production" in any sector.[113] Under these conditions, the introduction of any more capital only detracts from previous investments. It does so by diluting earnings per share on the one hand and, on the other, by increasing competition and therefore wages, temporarily further depressing the rate of profit as excess capital is used to pay workers above the socially necessary minimum.[114] This is the beginning of a process that culminates in debt crises, price reductions to move unsold goods, and interruptions in the supply chain as downstream vendors go unpaid.[115]

Ultimately, Marx asserts, excess capital must be sidelined or eliminated in one way or another for the economy to resume "'healthy'" operation.[116] In this sense, the "Law of the Tendential Fall in the Rate of Profit" is a financial rationale for capitalism's cycles of boom and bust, of overproduction and underconsumption, aptly summed up by the final sentence of the section of *Capital*'s third volume that bears its name: "Hence crises."[117] By contrast, the vision of full or nearly full automation favored by autonomists and accelerationists is one in which "the so-called law of value . . . is eroded and refuted by capitalist development itself" such that "crisis is no longer imputed to disequilibria which are linked to . . . the fall of the rate of profit."[118] These two visions of the unfolding of capitalist logic, then, and the modern schools of thought that derive from them, are generally opposed.

At the same time, as I have already pointed out, they both share an unwarranted optimism about the future of technological progress. It is thus crucial here to note that there is a significant stumbling block that precipitates an absolute overproduction of capital. This is the inability to technologically innovate. Reconsidering the declining rate of profit now from this perspective, 100% utilization of capital is also and simultaneously 100% utilization in the technological sense. "As soon as capital has grown in such proportion to the working population that neither the absolute labour-time that this working population supplies nor its relative surplus labour-time can be extended . . . there will be an absolute overproduction of capital."[119]

Where the length of the working day sets limits on absolute labor-time, it is machinery that determines relative surplus labor-time. It does so by allowing workers to do more, create more value, in the same amount of time and thus for the same wage. Absolute overproduction of capital thus describes a situation in which it is not possible to improve worker productivity by further investment in fixed capital.

Indeed, the need to address this accumulation by removing capital from the playing field "will also extend in part to the material substance of capital; i.e. part of the means of production, fixed and circulating capital, will not function and operate as capital, and a part of the productive effort that was begun will come to a halt."[120] The situation described here in theoretical terms is thus not far off from the scene Benanav paints of contemporary overaccumulation caused by overinvestment in fixed capital. "Overproduction of capital," Marx writes, "never means anything other than overproduction of means of production."[121] Yet capital still finds outlets for valorization within the industrial economy and other sectors, albeit at reduced rates, as well as in more speculative realms such as the stock market. The manufacturing situation has thus been more of a slowly percolating crisis than a turbulent upheaval, especially in comparison to financial crises.[122] For his part, Benanav admits that technical innovation not only continues in the manufacturing realm but that "in the context of intense global competition, high degrees of robotization translate into international competitive advantages, helping firms win larger shares of world markets for the goods they produce."[123] He rather contests the notion that these technological advances are widespread or innovative enough to have precipitated mass job loss. Instead, he argues, the decline in manufacturing profit rates disincentivizes both hiring in the sector and the introduction of transformative machinery.

For Marx, however, the solution to a crisis of overaccumulation is precisely such a revitalization of the means of production. An end to the turmoil brought about by the inability to valorize itself past a certain level of concentration comes about as

> the fall in prices and the competitive struggle . . . impel each capitalist to reduce the individual value of his total product below its general value by employing new machinery, new and improved methods of labour, and new forms of combination. That is, they impel him to raise the productivity of a given quantity of labour, to reduce the proportion of variable capital to constant and thereby to

> dismiss workers, in short to create an artificial surplus population. The devaluation of the elements of constant capital, moreover, itself involves a rise in the profit rate . . . With expanded conditions of production, a wider market and increased productivity, the same cycle of errors is pursued once more.[124]

The overall tendency of the rate of profit to decline is thus the outcome of a dialectical process in which overaccumulation and technological stagnation create high employment followed by crisis, while technological acceleration restores profits at a lower rate. In doing so, it diminishes the ratio between variable and constant inputs into the labor process, lowering the number of workers needed for a given task or quantity of output. The corresponding intensification of labor, given limits on the working day either imposed (collective organization for the eight-hour day) or immanent (the twenty-four/seven operation of digital platforms and services), is thus generally encountered today in "the accelerated speed of the machines."[125] This intensification exerts downward pressure on both wages and the cost of constant capital as more efficient production makes goods cheaper to produce.

The relationship between capital and technology here is quite clear: Overaccumulation would not be overaccumulation if capital could be productively invested in new machinery. Similarly, and more importantly for our purposes, the negative effects of overaccumulation cannot be overcome without a corresponding well of technological innovation to draw from. We have already encountered the argument that a lack of new computer upgrades would have an economic ripple effect not unlike those described by Marx as resulting from this type of convulsion.[126] Given that the initial conditions for a crisis in valorization stem from an inability to deploy more efficient means of production, these two scenarios are not just similar. Rather, 100% utilization figures as a key determinant in any postindustrial condition of absolute capital overaccumulation.

In outlining this cycle, however, Marx assumes the reservoir of capitalist technological revolution to be essentially limitless. So, too, do his heirs on both the accelerationist and automation-skeptical left. It is worth considering, however, the extent to which this modernist sentiment is a leftover from "the enormous development in the productive powers of social labour" of the mid-nineteenth century on the one hand, and its subsequent avatar in the form of Moore's law on the other.[127] Whichever side of the twenty-first-century debates on automation one may land on, to the extent that this repository of progress has not yet run dry, computing power has been a substantial reason why. It also offers one of the more significant counterexamples to the stagnation outlined by Benanav.

If there is one thing the pandemic demonstrated, it is that there has been little crisis of oversupply in the semiconductor industry.[128] After several years of this situation, recovery resulted in a temporary flooding of some sectors of the market as supply chains normalized. But even this effect is limited as long as next year's components are cheaper and more efficient than today's—that is, as long as Moore's law continues. At the same time, the pent-up demand from these shortages could not really spur further significant innovations. Research to improve semiconductor performance and the time required to build new fabrication plants taking advantage of it are already measured over an increasingly longer number of years. Benanav describes, correctly, the lack of economic innovation that a shift away from the manufacturing paradigm toward services represents, since services have proven much more difficult to scale than commodity production has. People have been pushed out of manufacturing and into more precarious labor markets, sometimes mediated through a smartphone app like Uber or TaskRabbit.[129] (Despite the supposedly measurable lack of computing's impact, it would nevertheless seem to be at least exacerbating this situation somehow.) However, there is also a parallel kind of precarity, which Benanav does not note, that is increasingly imposed upon integrated circuits. This is as sure a sign of stagnation for digital technologies as postindustrial labor force transformations are for workers.

Precarious Work, Precarious Labor

Over the past decade, security researchers have taken increasing note of hardware flaws introduced in efforts to increase microprocessor performance. This represents the downside of a development in which, to maintain their edge, processors began executing machine code in ways that were increasingly less obvious to those programming them in higher-level languages. As one such developer lamented, this was "a clear indicator of the extreme pressure the chip designers face to deliver ever-faster CPUs; they're under so much pressure that they'll risk changing the meaning of your program, and possibly break it, in order to make it run faster."[130] Vulnerabilities with names like Meltdown and Spectre take advantage of architectural changes that optimize the von Neumann bottleneck by computing and caching answers to problems before a program has determined they should be calculated—that is, by squeezing more computation into the idle cycles between operations and/or user input to ensure data is already available when these moments arrive.[131] Such techniques make computers do more for little additional reward, a concept with which anyone holding a job will be well familiar.

Were it possible to speak of the concept in relation to fixed capital, we could say here that an increased rate of exploitation leads to exploits. As Matt Spencer points out, these insecurities were anticipated and possibly even prompted by "Rowhammer" attacks exploiting similar efficiency hacks "to mitigate the effects of pushing chip architecture to ever smaller spatial scales" in the less heavily activated and therefore denser semiconductor configurations found in RAM.[132] Some of these vulnerabilities date back decades, but their identification in recent years demonstrates a growing awareness of the material exigencies of computation. Moreover, fixes—where possible—force manufacturers to sacrifice some of the performance advantages that inadvertently introduced the flaws to begin with, and this at a time when further gains in processing power are more difficult to come by.[133]

These issues can be contrasted against the Baudrillardian emphases of previous generations of vulnerabilities on symbolic metalepses. In doing so, we can index the increasing material—and not just conceptual—instability of computation in a Silicon Valley increasingly constrained by processor speeds. Traditional approaches to computer security are perhaps exemplified by UNIX coinventor Ken Thompson's 1983 Turing Award lecture. There, as a thought experiment Thompson proposed a compiler modified to insert not only a backdoor into any login program it compiles but also this same capability into any other compilers it subsequently compiles. Since any compiler built by the one so modified would itself produce compromised machine code even when working from known-good source code, the result would be a nearly indetectable exploit.[134] Hardware vulnerabilities focus not on these conceptual halls of mirrors but the material processes that they and every other computational operation depend upon.[135] In so doing, they highlight the compromises we introduce in our attempts to make our existing architectures do more and more. Just as we are unable to find a workable labor-market substitute for manufacturing, we have yet to devise a feasible postsilicon strategy for hardware, shunting people and computers alike into more precarious configurations.

These hardware and "wetware" situations increasingly mirror each other in being pushed, against seemingly intractable limits, to their breaking points. It is thus worth recalling them by their other names, introduced by Marx and expounded upon in their contemporary significance by Fredric Jameson: the forces and relations of production. As Jameson writes,

> These two terms of the opposition—material technology and human agency—stage two distinct perspectives—that of technological determinism and that of humanism—which are each of them clearly unsatisfactory . . . the insistence on relations, indeed, precludes that illusion of a purely technological determinism . . . and obliges us to dissolve the seemingly massive and impenetrable materiality of machinery back into its reality as human action (the introduction of new machines as a solution to the power of workers or to economic crisis, the very invention of technology itself, its human meaning as a specific form of exploitation or of discarding of redundant human beings).
>
> Thus this first opposition not only brings into visibility two distinct ways of reading or representing production itself: it posits each as the indispensable correction of the other, in a situation in which neither is the essential term in some asymmetrical opposition and which thereby demands a constant dialectical movement back and forth which must not be allowed to harden into a static sociological model of some kind.[136]

It is in this sense, I contend, that neither the electric dreams of the accelerationists nor the volitional transformation invoked by Benanav, who suggests that "instead of presupposing a fully automated economy and imagining the possibilities for a better and freer world created out of it, we could *begin* from a world of generalized human dignity, and then consider the technical changes needed to realize that world,"[137] are sufficient to address conditions of 100% utilization. Our situation appears more like an ever-tightening vise around computation and labor alike than the zero-sum imbalances with which dialectics are typically preoccupied. But this by no means indicates that they are not the operative mode of analysis here. We may instead be approaching something like a synthesis in which a long-held impetus toward computational determinism, inaugurated by the rise of discrete, electronic computing, is finally being confronted with the ultimate significance of the labor power it has tried to diminish.

As these two dynamics stumble toward an inflection point, the potential for—but, crucially, not the inevitability of—a radical restructuring between people and machines becomes stronger. It is of course impossible—if not outright wrong—to say for certain that there is no room for computational growth beyond Moore's law. Nevertheless, the current dynamics in technological and political economy alike are profoundly unstable and unsustainable. A situation that dramatically slows or even halts the replacement of labor with technology would provide an opportunity for exploited workers and their unemployed brethren to reassert the worthiness that Benanav suggests we take as a point of departure for a postcapitalist society. We have seen many indications that this could happen in the near future, from the physical impossibility of perpetual semiconductor miniaturization to the increasingly cruel optimization of really existing circuits, the unreliable embrace of neural networks and consequent diminishment of human capacities, and the faltering legitimacy of a political-economic order predicated on replacing people with microchips and justifying it with myths of meritocracy and endless exponential growth.

When such a scenario arrives, regardless of the specific circumstances or outcome, it cannot help but remind us how "labour . . . as the creator of use-values, as useful labour, is a condition of human existence which is independent of all forms of society; it is an eternal natural necessity which mediates the metabolism between man and nature, and therefore human life itself."[138] Here Marx delineates the limits of historicism by proffering a profoundly ahistorical concept. Any further consideration of the dynamic invoked by the limits of Moore's law, then, will therefore need to emphasize the difference between the transhistorical function of creative labor as the mediatic domain of human activity alone and the lossy thermodynamic transfer of energy known as work. Adapting a schematization already drawn by others,[139] efforts to move beyond this impasse should look to neither the folk songs of New Left solidarity nor the deterritorialized samples of drum'n'bass. They will need to take inspiration instead from the musical form arguably most preoccupied, since its inception in 1980s Detroit, with questions of automation and humanity—techno—and the words of one of its keenest twenty-first-century practitioners: *"Lasst . . . nicht die Roboter unsere Arbeit tun."*[140]

6 Beyond Homeostasis

What, then, is *our* work? The difficulty in conceptually disentangling it from that of machines—or, more fancifully, robots—is especially pronounced in German. There the word *Arbeit* is generally used to describe the activity of both people and machines, while *Werk* can refer to both individual feats and factories. So, too, is it in English, where the verb *to work* and the noun *work* are deployed with reference to both animate and inanimate subjects. Indeed, even the word *robot*, in its derivation from the Czech word for forced laborer, itself descended from the Old Slavonic for slave, illustrates this conundrum.[1] But English speakers also have recourse to the synonym *labor*, which generally describes only living—and predominantly human—effort. There is thus deep irony in the fact that, in their native language, the most vociferous defense of the value of human labor, in Marx, shares its vocabulary with that of physicists concerned with the technological and natural processes most suitable for replacing it. This linguistic situation showcases in miniature what is, at the level of capitalist societies, a much broader project to conflate these concepts. Their difference must therefore be investigated to conceptualize a world beyond 100% utilization—that is to say, a world where human labor can be understood as complementary, rather than subordinate, to computation. It is this distinction, between the capabilities of processing power, as an engine of physical work, and the capacity of labor to surpass it in crucial respects, that I want to suggest Villalobos's exhortation, in a language that does not as readily express it, is driving toward.

The human–machine interface is especially prominent in both the theory and practice of techno music. As its name suggests, the genre's deliberate foregrounding of technology immediately calls attention to the relationship between its means of production and the human beings responsible for producing it. Techno's origins in the automotive Rust Belt of the 1980s, just as the epicenter of the style and region alike, Detroit, found itself subject to increasing waves of both automation and outsourcing, make it a privileged artform for interrogating this association. As Benjamin Noys has argued, "Detroit techno was a self-consciously post-Fordist and post-industrial music. The very coldness of techno and its equivocal embrace of deindustrialisation make it discomforting."[2] Critical while also celebratory of the increasing prominence of computerization, this ambivalence reflects the "friction," as Noys puts it, "emerg[ing] from the tension of . . . utopian and futurist impulses with . . . moments of deceleration and anxiety."[3] Insofar as techno is a reaction to postindustrial developments, it is also a response to the impact of cybernetics, the scientific approach pioneered by Norbert Wiener dedicated to blurring distinctions between people and machinery.

Imaginary Solutions to Thermodynamic Problems

The cybernetic worldview treats both animate beings and inanimate objects as systems capable of intercommunication and mutual regulation. In so doing, it places living organisms and mechanical constructions on the same conceptual footing. This is clear from Wiener's influential introduction to the field, *Cybernetics; or, Control and Communication in the Animal and Machine*.[4] A key goal of cybernetics, in Wiener's formulation, was to stave off a system's inevitable loss of energy, resulting from the execution of physical work, through messaging and feedback between subsystems. The aim was homeostasis: a system whose continued operation, as much as and as long as possible, has been relieved of the tendency toward decay and entropy dictated by the second law of thermodynamics.

In the name of homeostasis, cybernetics sought to make the connection between machines and living beings as airtight as possible. An example offered by Wiener clearly illustrates the then-new discipline's dedication to finding similarities across the two:

> The conditions under which life, especially healthy life, can continue in the higher animals are quite narrow. A variation of one-half degree centigrade in the body temperature is generally a sign of illness, and a permanent variation of five degrees is scarcely consistent with life. . . . In short, our inner economy must contain an assembly of thermostats, automatic hydrogen-ion-concentration controls, governors, and the like, which would be adequate for a great chemical plant. These are what we know collectively as our homeostatic mechanism.[5]

The body, here, is akin to a factory, and the various self-regulating functions the equivalent of industrial apparatuses. Both are compelled to manage an "inner economy" ensuring their survival in the broader, "outer" environment. Cybernetic systems allow for a range within which operations can fluctuate, but pronounced disequilibrium—entropy—is strictly controlled through efficient means of communication. To the degree that these governing systems are organic, they can be learned from, emulated mechanically, and further controlled by such "artificial" methods.[6] To the extent they are technological, they can adapt biological processes to come up with new mechanisms for modulation and synchronization, including electronic methods of communication and calculation.[7] It is this conflation, rather than the more fantastical science-fiction visions, from which the concept of the cyborg or cybernetic organism arises: an interlinking of biological and machinic systems to ensure greater stability than either could achieve on its own. At the same time, the human was often found wanting in this calculus. As Fred Turner has put it, for cyberneticians, "underlying all their messy, fleshy, emotional complexity, human beings could be modeled as mechanical information processors. Moreover, if this was the case, they could be replaced by faster and more reliable mechanical devices."[8] In contrast to the slow, poorly understood pace of living beings, the growing speed at which machinery could compute information was a key driver of the cybernetic worldview and its real-world applications.

Attempts to erase the gap between work and labor by conflating the organic and inorganic have a much longer history, however. They stretch back to the onset of the industrial era, which gave rise to the science of thermodynamics. Chapter 3's discussion of custom silicon hardware has already touched on these concepts and their implications vis-à-vis the question of computational efficiency. However, their influence on cybernetics, hinted at in chapter 4, is equally significant. As economist and historian of science Philip Mirowski has argued, "the taproots of cybernetics and the theory of computation were buried deep in the loam of thermodynamics."[9] To the extent that the approach Wiener pioneered represents an extension of and response to these foundational physical principles, then, it illustrates a contiguity rather than breaks between an "industrial" nineteenth-, "postindustrial" mid-to-late twentieth-, and our early twenty-first-century capitalism threatened by 100% utilization.

As Anson Rabinbach has contended, the principle of the conservation of energy enshrined in the first law of thermodynamics and the second's articulation of a nevertheless inevitable tendency toward the dispersal of that energy as unusable entropy represented a fundamental antinomy of the industrial era.[10] Under this framework, he holds, the engine became the predominant metaphor for physical activity, one which itself collapsed the distinction between humans, animals, and machines as "the word *work* was universalized to include the expenditures of energy in all motors, animate as well as inanimate."[11] In contrast to cybernetics, and despite its close ties to statistical mechanics, this nineteenth-century equivalence was conceived largely in terms of physical force applied to tangible objects, as in heavy industry, rather than the informational and often microphysical realms that cybernetics would inhabit. Reinforcing the linguistic irony laid out at the beginning of this chapter, Helmholtz's term for the capacity to produce such work, coined in 1847, was *Arbeitskraft*, a term at once physical in nature and derived, at least according to a reflection by Helmholtz fifteen years later, via analogy with human labor.[12]

Rabinbach consistently translates *Arbeitskraft* as "labor power" when referring to either its human or mechanical manifestations, emphasizing how, in this way of thinking, "the energy of the violinist's body and the energy output of the industrial mill are identical: both represent labor power."[13] This certainly speaks to the conceptual basis by which nineteenth-century science and industry alike would seek to draw an equivalence between the two for the purpose of replacing the former with the latter. However, the appearance of the same term in Marx marks for Rabinbach a continuity where I would prefer to locate a rupture. In his telling, Marx's adoption of the principles of thermodynamics and the nomenclature of labor power/*Arbeitskraft* in his later works is an index of the theorist at his most accelerationist, gesturing toward "a future in which man will only have to stand watch over a self-regulating complex of technology."[14] At the same time, his identification of the falling rate of profit reveals "the principle of entropy at work in capitalism."[15] Marx, then, becomes a paragon of his age rather than a revolutionary seeking to transcend it. "In Marx's *Capital*, labor power appears as a force equivalent to other forces of production," Rabinbach writes, despite the fact that labor here possesses a quality that mechanical forces do not: the ability to create more value than its *Arbeitskraft* costs to reproduce.[16] Rabinbach argues that "Marx . . . viewed labor power (in contrast to labor) as devoid of purpose and meaning, a purely quantifiable output of force, subject only to abstraction."[17] Yet Marx himself describes "the capacity for labour [*Arbeitsvermögen*], in other words labour-power [*Arbeitskraft*]" rather differently: "a commodity whose use-value possesses the peculiar property of being a source of value."[18] In not distinguishing between work and labor-power, then, such a reading overlooks the extent to which Marx's formulation of not only *Arbeit* but *Arbeit*skraft worked *against* entropy. If labor's deployment in capitalist processes of valorization contributes to their decay, it is not through the exhaustion that is Rabinbach's focus but rather because labor transcends capitalism as a historically determinate regime of political economy.

Thermodynamic concepts of order and entropy were of paramount importance to Wiener's own thinking. However, he sought to stave off entropy not through labor and the eventual supersession of capitalism but rather via the governance of physical processes through systems, often electromechanical, of messaging and feedback. As he wrote in his popular account of the discipline,

> As entropy increases, the universe, and all closed systems in the universe, tend naturally to deteriorate and lose their distinctiveness, to move from the least to the most probable state, from a state of organization and differentiation in which distinctions and forms exist, to a state of chaos and sameness. In [Josiah Willard] Gibbs' universe order is least probable, chaos most probable. But while the universe as a whole, if indeed there is a whole universe, tends to run down, there are local enclaves whose direction seems opposed to that of the universe at large and in which there is a limited and temporary tendency for organization to increase. Life finds its home in some of these enclaves. It is with this point of view at its core that the new science of Cybernetics began its development.[19]

Built on the scientific turn toward probability in statistical mechanics that would eventually give rise to quantum physics, the cybernetic approach to regulating systems, as Orit Halpern makes clear, made peace with a limited amount of uncertainty in exchange for a greater degree of overall control. "Wiener was comfortable with acceding that the universe in its plurality," she writes, "might never be known. This accession, however, was only made to allow for the possibility that within far more localized situations, the future—chance—might yet be contained by way of technology."[20] If entropy could not be eliminated, it could, perhaps, be kept in check through high-speed information processing, allowing order, in some instances, to thrive.

Wiener's aim was to both replicate and enhance conditional tendencies for the rate of organization to rise. Cybernetics facilitates this through increased communication between various aspects of a system, viewed as a circuit of thermodynamic energy. It thus provides the link between a physical, industrial-era conception of work based on steam engines and an informational, ostensibly postindustrial landscape predicated upon computation. The heat problems that advanced integrated circuits now face can be traced directly back to this earlier framework. Moore's law does not supplant those of thermodynamics. In fact, the latter bear the formal status of scientific laws (however socially conditioned we may ultimately wish to view them) while the former holds its title only honorifically. The question thus becomes what to do when the negentropic force Moore's law represents, the local drive toward increased order and efficiency represented by growing computing power, nears the physical limits that ultimately constrain it.

Extending the connections we have already encountered between digital technologies, neoliberalism, and cultural stagnation, it is possible to view global capitalism itself as a kind of cybernetic—and in some sense, homeostatic—system. Despite Wiener's characterization of the idea "that free competition is itself a homeostatic process" as a "simple-minded theory," as scholars like Mirowski have shown, cybernetics, as "the philosophical overture to a whole phalanx of cyborg sciences," had a profound impact on postwar capitalist economics.[21] Viewed in this light, Wiener's declaration that "there is no homeostasis whatever . . . in the business cycles of boom and failure"[22] can be interpreted in one of two ways. It is either an ignorance of the metastability of this system of political economy despite these local turbulations, as diagnosed by Marx, or a particular sensitivity to, per Rabinbach, Marx's late-in-life theorization of capitalism's broader entropic trajectory.

Synthesizing Wiener's concerns with these cornerstones of Marxist dialectics, then, we can view capitalism as, paradoxically, a homeostatic system predicated upon continual growth. A metastable capitalism is one that is regularly expanding and whose periodic cycles of instability do not permanently upend the system but rather allow it to reconstitute itself at higher levels of accumulation and productivity.[23] Such self-regulation can be enhanced by "external" feedback mechanisms like monetary and fiscal policy.[24] One of its key *internal* regulators, however, is the decreasing ratio between necessary and surplus labor time enabled by progressively faster and more powerful machinery. Postindustrial capitalism has already lost much of its dynamism, as the critiques of stagnation we have previously encountered at both the level of culture and economics have demonstrated. This illustrates, again, the extent to which relative rates of improvement, which are faltering, are as important if not more so than absolute accumulation, which continues to increase.

Growing chaos both online and off may very well be one sign of how these slowdowns are leading to a larger breakdown. In the absence of sustainable growth, politics becomes a more palpable contest over the diminishing remainder. Yet an end to neoliberal homeostasis could have potentially positive effects as well. A collapse in the existing order need not result in chaos if it is replaced by a new and more equitable one—that is, if its reconstitution takes the form not of a more extensive capitalism but something else. In his antihumanist accelerationist reading of Bitcoin, which attempts to render Marxism dependent on thermodynamics, cybernetics, and even Moore's law itself (rather than seeing it as transcending them, as I am arguing here), Mark Alizart nevertheless offers the following observation: "Human work"—for the reasons outlined earlier I must insist on the term labor—"is capable of 'negentropy.'"[25] In contradistinction to a decaying regime which has sought to minimize it, labor, in its capacity to act as a creative and thus negentropic force, can function as a basis for new ways of organizing not only ourselves, but our relationship to technology in a postcapitalist, post–Moore's law world.

Postindustrial Lies

Detroit techno's ambivalence over the kind of human–machine symbiosis implied in a theory of cybernetics can already be found in one of its key influences. The very name of the band Kraftwerk, German for "power plant," already indicates a deep connection with thermodynamics and, by extension, the questions of energy, work, and labor we are concerned with here. Befitting their status as pioneers, it is useful to think of the group as straddling the conceptual line between classical thermodynamics and cybernetics, whereas its techno protégées fall more straightforwardly into the latter camp. The closing, title track from Kraftwerk's 1978 album *The Man–Machine* plays ambiguously with its conceit. Is the ensemble it describes a "pseudo-human being" or a "superhuman being"?[26] The German lyrics instead lean more heavily into the technological side of the equation, alternating instead between "half being and half thing" ("*Halb Wesen und halb Ding*") and "half being and half super-thing" ("*Halb Wesen und halb über Ding*").[27] In both cases, the song's use of a vocoder to synthesize its lyrics with a nonhuman sound source reinforces this concept.

Famously, the vocoder was used during the Second World War, the crucible of cybernetic and information-theoretical development, to prepare human speech for digitization. The device separated the voice into frequency bands whose outputs could be assigned a number, which could then be interpolated with a similarly digitized source of random sound in order to produce an unbreakable signal for radio transmission.[28] This system, known as SIGSALY, X-System, or Green Hornet, "was the first practical example of digital speech transmission," and thus a key example of the adaptability for which the nascent mediatic regime of electronic digital processing would come to be known.[29] Developed in the First World War and understood to be completely unbreakable, the technique of using a truly random key as long as the message, alongside the injunction that it never be reused, was known as a Vernam cipher.[30] This method was an elaborate version of simple substitution ciphers dating at least as far back as the Roman Empire, "the basis," for Friedrich Kittler, "on which command, code, and communications technology coincided."[31] Specifically developed for alphabetic messages sent over the telegraph or radio, however, the Vernam cipher is in many ways a product of modernity despite these classical roots. The strict and unwieldy requirements of the system, Vernam wrote, "are such as to make it unsuitable for general use, unless mechanical methods are used."[32] In rendering auditory information encipherable in the same way, both Claude Shannon and Alan Turing consulted on the SIGSALY project in various capacities, another way that their theories and practices ushered in a postmodern era in which all manner of signs and symbols became interchangeable, computable information.[33] Indeed, as one of the earliest efforts at discrete, electronic information processing, the X-System represents a turning point in the trajectory toward what Kittler argues is a new "world empire" marked by "the coincidence of communication technologies and Turing machines."[34]

Over the course of World War II, Shannon was able to mathematically prove the Vernam cipher's security. Afterward, this demonstration was published as part of a follow-up to his foundational article introducing the concept of information theory, illustrating some of its practical cryptographic applications.[35] Shannon's proof shows that the entropy of the enciphering key, in the informational sense of "a measure of how much 'choice' is involved in the selection," is the determining factor in securing a message such that not even the most powerful computers could break its code.[36] As Turing's biographer Andrew Hodges puts it, "If all keys were equally likely, then no weight of evidence could accrue to any particular possible plaintext."[37] A truly random, never-reused key as long as the message it enciphers makes the latter impenetrable because any decryption is as probable as any other. Any four-letter message could be MOVE *or* STOP, for example; without the key it would be impossible to know which.

Here it is not a matter of processing power but rather the equanimity of chaos: There is simply no way to determine the original message without its key. This fundamentally aleatory quality distinguishes the Vernam cipher from most other cryptographic systems, which strike a more delicate balance between determinacy and entropy. Earlier ciphers using a shorter, repeating key were vulnerable to attacks based on the frequency of recurring letters. Systems from the late 1970s on, on the other hand, typically do not require a preshared, unresuable key, making them suitable for things like electronic commerce. However, they are also susceptible to decryption if enough computing power is brought to bear upon them. In short, both types have a verifiably "correct" key such that a cryptanalyst will be able to tell when they have successfully deciphered a message. Vernam-encoded messages offer no such reassurance.

In the case of SIGSALY, thermal noise from gas-filled electron tubes recorded onto vinyl records provided the necessary randomness.[38] The system thus placed entropy, in not only the informational but the thermodynamic sense, in service of order. In this way, it was a cybernetic system par excellence. Provided the proper security measures were maintained, the homeostatic integrity of the plaintext was guaranteed to be accessible by sender and receiver alone, regardless of the quantity of calculative power thrown at it. In an inversion of the Derridean concept that the statement or mark is always constituted by that which undercuts it as parasite (also French for noise),[39] organization was buried under decay, yet always ready to be reanimated.

Within this system, the vocoder allowed audio to be subject to the same rules as written letters, specifically, substitution. Once in this state, the relatively simple mathematics necessary for Vernam encryption could be automatically performed on the human voice. It was, quite literally, a "man–machine interface." Shannon's work, including on the vocoder itself, is referenced repeatedly by Wiener, who notes that the technical requirements of signal encoding and transmission

> must be supplemented by the use of decoding devices adequate to transforming the received information into a form suitable for reception by human receptors or use by mechanical receptors. Similarly, the original message must be coded for the greatest compression in transmission. This problem has been attacked, at least in part, in the design of the "Vocoder" system by the Bell Telephone Laboratories, and the relevant general theory has been presented in a very satisfactory form by Dr. C. Shannon of those laboratories.[40]

Again, here, we encounter an effort to equate the functions of humans and machines: both have "receptors" for processing information, according to which it must be (re)formatted. The vocoder can thus itself easily be viewed as a cybernetic device. It translates between and therefore brings into parity the mechanical and the organic at the level of intercommunication rather than in terms of their capacity to perform the same type of physical work.

Moreover, though in SIGSALY it was not the vocoder but the digital parts of the system performing the encipherment and decipherment, adding and subtracting the key and plaintext bitstreams modulo six to do so,[41] Wiener argued that analog audio technologies could be understood as types of computers in their own right. "Telephone receivers, wave filters, [and] automatic sound-coding devices like the Vocoder," he writes by way of introducing information theory, "are all in essence quick-acting arithmetical devices, corresponding to the whole apparatus of computing machines and schedules. The ingenuity needed in their use has been built into them in advance. . . . The chain of operation has to work too fast to admit of any human links."[42] In cybernetics and early information theory, computation is the means by which human expression—in this case, speech—and machinic automation can be productively coupled. In doing so, it increasingly places the human at a remove from any intermediary steps in the process. Processing speed thus becomes a relevant metric not only for the internal aspects of the apparatus but its human-facing elements. These must treat the functionality of biological sense organs as a set of specifications for receiving and presenting data, and handle this information in a timely fashion.

This sense of the vocoder as essentially cybernetic and evocative of computation despite its analog provenance remained, as Dave Tompkins notes, long after it was unpaired from digital cryptography and became a popular musical instrument. In contrast to both visions of a technological dystopia and the secretive sessions that produced *The Man–Machine*, he writes, the German company

> Siemens held a sunnier outlook on the relationship between humans and computers. A 1973 press release for the vocoder describes the computer as "a conversational companion" and follows that "communication between man and machine is a desirable goal." The German rendition of "man and machine" appears on the press release as "Maschine–Mensch," replacing conjunction with hyphen and allowing robots to make the band, one and the same.[43]

This coupling—and the ambivalence with which Kraftwerk, if not equipment manufacturers, would approach it—would be picked up by what is generally recognized as the first Detroit techno group, Cybotron. The band began as a collaboration between a young Juan Atkins, who would subsequently become one the founding figures of Detroit techno, and Rik Davis, a Vietnam veteran also going by the alias 3070. Its name combines the concepts of the cybernetic organism, or cyborg, with the cyclotron: the ostensibly positive, homeostatic interrelationship between people and technology and the high-speed apparatus associated with particle physics.[44] Alternately, as Tompkins puts it, "a particle accelerator and the type of robot that puts humans out of business."[45] The cover of the group's debut album, *Enter*, depicts a man in the process of running through a virtual window. Half in and half out, the right side of his body dematerializes into a blur of scanlines. On the back, the same figure appears, now having fully crossed the electronic Rubicon, as a posthuman ghost.[46]

Music critic Jon Savage argues in the liner notes to a compilation of the band's work, tellingly titled *Interface: The Roots of Techno*, that "if there is one central idea to Techno, it's of harmony between man and machine."[47] Yet we should take pains not to arrive at this cybernetic promised land too early. As we have already seen Noys suggest, the genre is exceedingly aware of the complications that can result from this pairing. Though he concludes by celebrating "the positivism inherent in techno," Savage also notes that the group's 1984 single, "Techno City," which arguably gave the genre its name, is "precisely balanced between utopia and dystopia, between the future as dream and the future as nightmare."[48] *Enter*, released the previous year, makes this ambivalence clear from the start.

A document of an artform in chrysalis, the album sits between the dancefloor-oriented genre techno would become (in no small part due to Atkins's guidance) and the earlier, cerebral, forms of progressive rock that inspired Davis and drew him to dystopian science-fiction narratives. *Enter* is also explicit in its engagement with the politics of its time. "The Line" describes the experience of unemployment, one that at least implicitly, given the band's name, musical style, and temporo-spatial location, evokes the specter of automation. "Industrial Lies," on the other hand, offers a broader critique of political economy. The album's penultimate track, "El Salvador," updates Davis's war experiences via reference to 1980s conflicts. Its "lyrics . . . only understood by Davis himself, who was then processing some post-traumatic demons through a Korg vocoder," the man–machine interface returns to the battlefield to technologically encipher the too-human experience of war.[49]

On the other hand, Davis also champions "the necessity of interfacing the spirituality of human beings into the cybernetic matrix."[50] The album's closer and undoubtedly most well-known song, "Clear," similarly provides a template for the future, and futuristic, direction the music would take. Viewed darkly by Davis, it nevertheless offered a danceable, "technocratic vision," more closely associated with Atkins and indebted to Alvin Toffler's *The Third Wave*: "Could machines replace human assembly lines so GM could chew up the highways that had already destroyed black neighborhoods?"[51] Similarly, music writer Simon Reynolds quotes Atkins's assertions that "the 'double-edged sword' of technology" was central to the group: "'With technology, there's a lot of good things, but by the same token, it enables the powers that be to have more control.'"[52] Reynolds further argues that "for all their futuristic *mise-en-scène*, the vision underlying Cybotron songs was Detroit-specific, capturing a city in transition: from industrial boomtown to post-Fordist wasteland, from US capital of auto manufacturing to US capital of homicide."[53] From Kraftwerk to Cybotron to Villalobos, the lineage of techno is not simply one of cybernetic synthesis but an ambivalence about the possibilities and realities of what such a technological convergence could look like.

Noys argues that techno's

> invention lay in the welcoming of the "mechanization," or better "computerization" of the aesthetic (which had obviously been prefigured by Kraftwerk's *Man–Machine* (1978) and *Computer World* (1981)) . . . the equivocations lay in a sense of abandonment: an escape to the future, escape from labor, or the loss of labor and the collapse of the future into permanent unemployment?[54]

Noys's position, though not necessarily his approach, resonates with the critques of accelerationism we have encountered previously. Seeing the tendency and its influences from Italian Futurism to Thomas Pynchon and the CCRU as offering an ideological vision of technological progress that sits uneasily with a stagnating postindustrial political economy, he ultimately argues that "we have to tap and resist the incitement of desire that capitalism produces and which accelerationism mimics—the fantasy of immersion into Real forces of acceleration."[55] It is thus significant that Noys not only acknowledges the ambivalence within the genre but expresses his own. If, on the one hand, "Detroit techno seemed to celebrate the replacement of 'variable capital' (i.e. humans) by 'constant capital' (i.e. machines) at Ford," on the other it "could be re-read . . . as a critique of the 'smoothness' of acceleration, by a repetition that disrupts the future rather than the endorsement of accelerationism."[56] To not let the robots do our work, then, could be seen as an example of this latter rupture, a rethinking of the man–machine interface around some alternative and potentially noncybernetic vision.

At the same time, Noys's critique in a sense admits too much, particularly around the question of processing power. Arguing that "the computer . . . embodies the 'speed-up' of labor, as each new model becomes faster and faster (or that is the promise)," he also argues that this embodiment is essentially false due to the productivity paradox. "A certain attachment to an accelerative dynamic of 'productive forces' . . . proved illusory," he writes. "Accelerationism enhances and celebrates this, but the future it could not grasp was the future of crash and crisis—the terminus of acceleration in the grinding to a halt of the speed machine of capitalism."[57] It is noteworthy that it is capitalism, and capitalism alone, that is predicted to seize up here. Indeed, despite the tantalizing parenthetical quoted above, Noys's only real engagement with the physical limits of computation comes in an offhand acknowledgment that high-frequency trading is coming up against them, as he takes Nick Srnicek and Alex Williams to task for refusing to endorse massive processing power as an example of the type of accelerationism they champion.[58]

By contrast, for Noys, techno, cyberpunk literature, and the CCRU rely on "not only the logic of increased computing speed and power, but also the claim that capitalism is maintaining its dynamic of acceleration first given its most memorable form by Marx and Engels in 'The Communist Manifesto.'"[59] The implication here is clear: Computation *has* continually increased in speed and power, but this does not matter economically. It therefore can stand in, in an ideological fashion, for a lack of economic dynamism identified by late twentieth- and early twenty-first-century Marxists. We mistake real, but irrelevant, increases in computing speed for the real, and meaningful, decline in profit rate.

If, however, the mistake of accelerationism is not this conflation of technology with economics but rather its unwarranted assumption of continual improvement in computational capacity, the calculus changes. The question of homeostasis would return, with increases in processor efficiency working to regulate stagnation or even decline in other segments of the economy by incorporating them into the realm of data processing. Computation's own terminus, one that neither accelerationism nor its critics take into account, could then destabilize the delicate equilibrium of late capitalism. It is under this set of circumstances—100% utilization—that distinguishing between human labor and machinic work becomes critical, and under which techno's perspective on their postwar interrelationship takes on a different valence.

Work and Labor After Cybernetics

A similarly nuanced approach, though ultimately more celebratory than techno's, to the kinds of interlinking that cybernetics enables can be found in the technoscientific, ecofriendly, socialist-feminist theory of Donna Haraway. For her, the cyborg arises out of the conceptual conflation of humans, animals, and machines (though as we have seen this was already a part of nineteenth-century thermodynamics). On the one hand, "the cyborg appears in myth precisely where the boundary between human and animal is transgressed," while, on the other, "late twentieth-century machines have made thoroughly ambiguous the difference between natural and artificial, mind and body, self-developing and externally designed, and many other distinctions that used to apply to organisms and machines. Our machines are disturbingly lively, and we ourselves frighteningly inert."[60] The cyborg's rise to prominence out of Western Cold War institutions holds the promise, for her, of destabilizing a vast array of historical boundaries and binary oppositions, from those between living and nonliving things to those between genders, identities, and, importantly, what counts as work.

In this view, feminist contributions to the discourse around labor that expand it outside of the factory and into the domestic and reproductive sphere (labor, of course, also being a word for the process of birthing a child) are significant, but do not go far enough. "The home, workplace, market, public arena, the body itself—all can be dispersed and interfaced in nearly infinite, polymorphous ways, with large consequences for women and others—consequences that themselves are very different for different people and which make potent oppositional international movements difficult to imagine and essential for survival," she writes.[61] Notably, she is less skeptical of "the essentializing move . . . in the ontological structure of labour or of its analogue, women's activity" than "the inheritance of Marxian humanism, with its pre-eminently Western self."[62] Indeed, she sees postindustrial labor as "redefined as both literally female and feminized, whether performed by men or women . . . able to be disassembled, reassembled, exploited as a reserve labour force,"[63] as well as characterized by some of the racial dynamics explored here previously. "Black women in the United States," she notes, "have long known what it looks like to face the structural underemployment ('feminization') of black men, as well as their own highly vulnerable position in the wage economy."[64] Haraway's explicit championing of the labor of women in Asian electronics assembly plants is a key influence on Lisa Nakamura's analysis of similar employment in Native American reservations also considered in the last chapter. It is the figure of the cyborg, as a figure of hybridity, that can construct a new kind of solidarity out of these disparate oppressions and their myriad intersections: "The cyborg is a kind of disassembled and reassembled, postmodern collective and personal self. This is the self feminists must code."[65]

If the cyborg is a complex of intermediations that enables the articulation of various distinctions without erasing the uniqueness of its individual parts, however, it is also one that is enabled in large part not just by computing power, as this reference to coding implies, but specifically by Moore's law. Alongside the blurring of boundaries between living and inanimate matter, in a passage worth quoting at length, Haraway argues that her perspective is enabled by a collapse of the distinction between the material and the immaterial enabled by the shrinking size of transistors.

> Modern machinery is an irreverent upstart god, mocking the Father's ubiquity and spirituality. The silicon chip is a surface for writing; it is etched in molecular scales disturbed only by atomic noise, the ultimate interference for nuclear scores. Writing, power, and technology are old partners in Western stories of the origin of civilization, but miniaturization has changed our experience of mechanism. Miniaturization has turned out to be about power; small is not so much beautiful as pre-eminently dangerous, as in cruise missiles. Contrast the TV sets of the 1950s or the news cameras of the 1970s with the TV wrist bands or hand-sized video cameras now advertised. Our best machines are made of sunshine; they are all light and clean because they are nothing but signals, electromagnetic waves, a section of a spectrum, and these machines are eminently portable, mobile—a matter of immense human pain in Detroit and Singapore. People are nowhere near so fluid, being both material and opaque. Cyborgs are ether, quintessence.[66]

Integrated circuitry here is celebrated for its ostensible immateriality in its role as mediator of both digital and analog signals. Despite being based on writing, understood as an imperialist technology of oppression, microchips subvert this historical trajectory, upending patriarchy and logocentrism by operating at unfathomable scales and holding the potential to "subvert the apocalypse of returning to nuclear dust in the manic compulsion to name the Enemy."[67] The diminishing size of electronic components enables them to transcend the national borders that still constrain larger entities like people, giving rise to cultural expressions of this new paradigm, such as Detroit techno. By the same token, to the extent that cyborgs are not simply machines but hybrid assemblages of humans and animals as well, the latter should aspire to this level of ethereal celerity in their merger with technology.

Haraway's understanding of the relationship between the history of such advanced technologies, stemming from the heights of Western rationality, and their potential to subvert many of its most well-known precepts, is complex and nuanced. "Illegitimate offspring," she writes, "are often exceedingly unfaithful to their origins. Their fathers, after all, are inessential."[68] Similarly, she counts "technological determination" as "only one ideological space opened up by the reconceptions of machine and organism" she charts.[69] Though she would surely reject this characterization, there is something of the dialectical in her perspective here, which echoes both Marx and Atkins in its ability to acknowledge both the rewards and risks its object of analysis holds:

> From one perspective, a cyborg world is about the final imposition of a grid of control on the planet. . . . From another perspective, a cyborg world might be about lived social and bodily realities in which people are not afraid of their joint kinship with animals and machines, not afraid of permanently partial identities and contradictory standpoints. The political struggle is to see from both perspectives at once because each reveals both dominations and possibilities unimaginable from the other vantage point.[70]

This underlying motif is also visible in another rhetorical move she makes. As a paragon of hybridity, "the cyborg has no origin story in the Western sense—a 'final' irony since the cyborg is also the awful apocalyptic *telos* of the 'West's' escalating dominations of abstract individuation, an ultimate self untied at last from all dependency, a man in space."[71] Those who seek to reject teleological thinking both on the right and the left often make similar arguments in a kind of intellectual jiujitsu whereby the weight and momentum of History is leveraged into passing a point beyond which it can no longer stand.[72] Dialectics and/or teleology, the claim goes, has brought us this far, but can go no further. Responding to Haraway's statement, Mirowski makes a similar observation: "This saga of the development of a world-historical theme into its antithesis may smack of the Hegelian dialectic—a *genre* that has never entirely lost its appeal, no matter how despised and disparaged—but it is also something else: it illustrates the ease with which machines undergo transubstantiation into thoughts, as well as vice versa."[73] Here, "the certainty of what counts as nature," and thus any reconciliation of nature and consciousness, "is undermined, probably fatally" by the teleological arrival of the cyborg.[74] The real and the rational become indistinguishable but in some paradoxically non-Hegelian fashion.

Haraway's text has been massively influential in its articulation of a postindustrial technological base with a range of progressive cultural and political movements and objectives. These latter goals and tactics—the need to construct an assemblage of the oppressed that cuts across without diminishing differences of gender, race, and types of labor, for instance, or the imperative to forge less exploitative relations with nature—are important, and as-yet only very partially realized, ones. At the same time, as Fenwick McKelvey and Joshua Neves have argued, with the arrival of the "cyborg sciences . . . there is a real tragedy . . . as pacifists, socialists, and humanitarians turned toward optimization as a way out of the insanity of nuclear war or capitalism only to find new articulations of power."[75] The challenge and opportunity for such lines of thought in an era of 100% utilization, then, is to unshackle these social ambitions from a technological current that is now finding itself unable to move them forward and the power relations accompanying it.

One place to begin is in the realization that "the boundary-maintaining images of base and superstructure . . . or material and ideal" which, under this new technosocial order, supposedly "never seemed more feeble"[76] have in actuality *gained* strength on the basis of the microelectronics revolution enabled by Moore's law. What appeared as a collapsing of these binaries was actually enabled by the increasingly rapid capacity of transistors to switch between two states, and all the implications that has for the kind of higher-order computational operations that might support the cybernetic organism, human–machine interface, or virtual worlds, the vitality of the machines in whose countenance we now appear motionless. But if dialectics have not yet come to a halt and a kind of teleology persists in the form of growing computing power, its terminus is at the very least hinted, if not indicated, by the end of Moore's law. It is thus up to us to rescue these more political aims from their association with the performance of the microprocessor.

As Haraway implies, foregrounding labor, denuded from its associations with Western humanism (including that of the Marxist variety) is one possible answer to this conundrum, because "Marxist/socialist-feminism does not 'naturalize' unity; it is a possible achievement based on a possible standpoint rooted in social relations."[77] Certainly this concept is already enshrined in Marx's concept of socially necessary labor-time. Machinery plays a key role, as what we may think of at any given moment as labor can abruptly lose its theoretically superior status as such if it becomes commonly achievable by machines. After "the introduction of power-looms into England," Marx argues, "the English hand-loom weaver in fact needed the same amount of labour-time as before; but the product of his individual hour of labour now only represented half an hour of social labour, and consequently fell to one half of its former value," as the machine-aided laborer could achieve the same output in half the time.[78] The social status of labor, and thus the possibility for collective action, is determined in conjunction with its relationship to machines.

In a reading of an obscure text authored by one R. S. Hunt and deposited in Wiener's collected papers as a result of the former's efforts to secure an audience with the latter, Seb Franklin builds on Rabinbach's analysis and others to explore the implications that cybernetics and information theory hold for the postindustrial redefinition of labor. Hunt's contribution, as Franklin makes clear, was the concept of "'G' . . . a form of energy that passes from human beings, nonhuman animals, and machinery into 'material systems.' In contrast to what Hunt defines as A-energy—that which takes the form of heat, light, motion, and so on and is subject to the laws of thermodynamics—G-energy transforms a material system 'from a more to a less probable condition.'"[79] While, as we have already seen from Wiener, these two aspects are not contrasting in that Hunt's "A-energy" *also* transforms systems from high-probability to low-probability states,[80] where the *real* distinction lies is in the latter's purported imperviousness to thermodynamic laws. G-energy "draws on information theory to undo the distinction between value-producing labor and phenomena such as will or moral purpose, which in [Hunt's] theorization belong to an entirely different category of energy,"[81] a conceptual innovation which Franklin identifies as a precursor not only to a shifting understanding of labor in postmodernity but also autonomist critiques of it.

What is at stake here is precisely the defining feature of computation as automation, the way that it can replace ostensibly physical processes with ostensibly symbolic ones.[82] Cybernetics extends the thermodynamic equivalence of animals, machines, and people from the realm of physical work to intellectual and affective labor. If all labor can then be reconceived as "immaterial labor"—that is, if it can be understood as an effective procedure—Franklin argues, then Hunt's "conceptual shift retroactively defines precomputational forms of labor in terms of some underlying process of informatic exchange"[83] impervious to entropic decay. For Franklin, the implications of this line of thought are both fantasmatic and illustrative of a contiguity between industrial and informational capitalism. "The materially deleterious return of fantasies of unbounded accumulation through principles of information-without-energy" demonstrates "how the social relations of capital in both 'classical' and 'post-industrial' forms are connected by the fantasy of value as a material essence that can be found in and extracted from social bodies *without any material cost to those bodies.*"[84] G-energy thus represents "a wholesale reformulation of the relationship between living labor and dead labor."[85] Simultaneously, Franklin argues, it participates in an ideological project to sustain accumulation, against the reality of a declining rate of profit, through an expansion of what counts—and what can be counted—as labor.

There is much to recommend this analysis, which emphasizes both the social characteristic of labor as well as the toll it can take on the people who carry it out. In these very emphases, however, larger questions about the nature of labor, as well as the nature of machines, can be elided. To my mind the fundamental error in Hunt's thinking, one surely noted but not fully explored by Franklin, is not in the notion that physical procedures can be reformulated as informational ones, but rather that the latter somehow operate at no thermodynamic cost. As we have seen throughout this book, this is simply untrue, and this fact has implications for machinery as profound if not more so than it does for humans. Furthermore, Franklin's emphasis on the "material cost" to "social bodies" threatens to diminish the very distinction between living labor and thermodynamic work that renders the former constitutively irreducible to the latter. However, if so-called informational work *were* subject to the same principles of energetic transfer as mechanical systems—which it most certainly is, because informational systems are material configurations operating in the same physical universe—then the most unique quality of labor would no longer be threatened by this computational subsumption.

Far from upending the distinction, in enabling a wider range of activities to be carried out by machines computerization follows in a much longer history of automation that renders what had previously been considered living labor "dead." In Marx's analysis, capitalist production already takes entropy into account: the physical deformation of tools and machinery utilized in the valorization process is transferred into the value of the commodities they produce, but does not generate *new* value.[86] As with the transfer of energy in thermodynamics, machinery "never adds more value than it loses, on an average, by depreciation."[87] As we have seen, once relegated to machines, thermodynamic work can add value only under the very specific circumstance in which the labor carrying it out remains cheaper than machinery to employ. In fact, Marx condemns this condition as both paradoxically encouraged by the intensification of structural unemployment through widespread industrial adoption of machinery and a "shameless squandering of human labour-power for despicable purposes."[88] The informationalization of labor processes simply increases the number of functions subsumable under the category of thermodynamic work insofar as they are capable of being carried out by discrete, electronic computers but not industrial machinery.

The question of what it means to *not* shamelessly squander labor-power thus arises. Here squandering does not mean the appropriation of surplus value and the like but rather the continued employment of humans at subsistence wages for tasks that could be performed by machines. This certainly militates for Marx as a proponent of efficiency. As Rabinbach puts it, "Marx's most eloquent praise of capitalism is reserved for its revolution in production, just as his most hostile remarks are reserved, not simply for its unjust and exploitive aspects, but for its 'anarchic system of competition,' which wastes the energies of society: 'the most outrageous squandering of labor power and of the social means of production.'"[89] Here the section Rabinbach quotes, however, reinforces not the commitment to the *elimination* of labor he wants to see in Marx but rather the necessity of its *transformation* in a postcapitalist society:

> Only the abolition of the capitalist form of production would permit the reduction of the working day to the necessary labour-time. But even in that case the latter would expand to take up more of the day, and for two reasons: first, because the worker's conditions of life would improve, and his aspirations become greater, and second, because a part of what is now surplus labour would then count as necessary labour, namely the labour which is necessary for the formation of a social fund for reserve and accumulation. . . . The absolute minimum limit to the shortening of the working day is, from this point of view, the universality [*Allgemeinheit*] of labour.[90]

Labor is not abolished under communism, but rather, in the latest iteration of its transhistorical nature, radically refashioned into a collectively organized activity. Similarly, Franklin is concerned with other kinds of labor overlooked in "the fantasy of labor as an essence that can be separated from the supposedly nonproductive activities of reproduction—principally those racialized and gendered forms of biological reproduction and care that take place in diverse locations and contexts, and can be both waged and unwaged, but are always devalued, occluded, and abjected."[91] Yet while labor may be inseparable from reproduction (indeed, as the very word suggests), it is also not reducible to it. It is here that the discourse of exhaustion Rabinbach, Franklin, and others emphasize risks foregrounding the thermodynamic qualities of the human organism rather than the negentropic qualities of labor as such.

Labor-power is not squandered precisely when it is *applied toward a task that machines are incapable of completing*. Indeed, if we accept Helmholtz's post hoc ascription of his use of *Arbeitskraft* to an analogy with human labor, this may have been the case all along. As Rabinbach writes, "For Helmholtz . . . human skill is valuable only where it cannot be replaced by machines."[92] Though unremarked upon further, this formulation does not simply emphasize the social quality of labor and retain a distinction between it and work. At another level, it acknowledges their physical equality as thermodynamic mediations, as the activity of things that direct energy and thus inevitably decompose. In this latter view, as Rabinbach writes, "the only essential difference between natural forces, machines, and human beings is in the utility and efficiency of the energy transferred."[93] Yet however labor may be exploited, or even unrecognized, it represents a creation of value that machines alone cannot provide.

Importantly, this generative capacity exists in the laborer *alongside* the more thermodynamic quality of transmission. It is, for Marx, "the twofold nature of his [*sic*] labour: it must at the same time create value . . . and preserve or transfer value" from machinery and raw materials.[94] Nevertheless, the former activity is a property of *labor*, not machines. It is on this basis that a claim can be made for its uniqueness vis-à-vis the universal physical laws that govern *work*. In creating more value than it costs to reproduce, labor functions as the loophole through which capital valorizes itself, how money becomes more money through the mediating form of investment—that is, capital. It is also, at the same time and dialectically, the capacity for transcending this mediation. These two axioms form the foundation of the first volume of *Capital*. In cybernetic terms, labor is a negentropic function capable of building both a world of material abundance and a society predicated on something other than the property relation. More importantly, it is a kind of negentropy that can transcend homeostasis rather than merely preventing disorder. In this sense, it is anticybernetic.

Returning finally to the question of 100% utilization, it is precisely such a possibility that a crisis in semiconductor development brings to the fore. If a computational process is effectively prevented from standing in for human activity due to the incapacity of hardware to execute it in a timely fashion, then the further development of automation is essentially held hostage by the laws of physics. The encroachment of fixed capital upon labor processes is stalled by the material operation of the machinery upon which any replacement procedure must execute. The problem we are faced with here is not physical heat death so much as a breakdown, at the level of political economy, in the delicate balance between human and machinic inputs.

As any hardcore gamer will tell you, given proper cooling, a computer running at full utilization can still operate, and potentially for an extended period of time. Further, should it break down, it can easily be replaced with another one. Computers are themselves commodities that increasingly operate according to a logic of interchangeability that will only be intensified to the extent that their technical development ceases. The true homeostatic threat, rather, is to a mode of production reliant upon an ever-increasing range of activity being subsumed into mechanical processes, a category within which we now have to admit informational procedures. Computation is not immune to the laws of thermodynamics. It does not represent any kind of break from industrial processes but rather a contiguity masked by its pursuit of the upper bounds of efficiency, now seemingly approaching a limit. As this happens, labor increasingly becomes visible in the remaining human faculties unassimilable to machinic computation. This situation should not be confused with perennial, and therefore often ahistorical, efforts to call attention to labor wherever it exists in the world. It is rather only through an understanding of the cutting edge of capitalist production, of what it means to automate human activity away, that we can come to understand what labor really is. More importantly, it is the only way we can uncover the potential labor holds.

A similar concept in a twentieth-century, antihumanist vein can be found in Friedrich Kittler's references to "so-called man," the residue of human chauvinism left over after each successive wave of technical media emulates some activity or capability previously understood as the exclusive domain of people. As his translators write, "We are faced with the *Aufhebung* of human processes into silicon microprocessors, the dissolution of human software into computer hardware, for if computer technologies . . . ultimately substitute for physiological impairments and extend the sensory apparatus, then technology's prosthetic function could allow for the complete replacement of the human."[95] It is thus notable that Kittler himself would subsequently allow for a way out from this trajectory. As we saw at the beginning of this book, the physical limits to silicon hardware form the basis for a grudging admission that "the familiar face of man" may not in fact be entirely eclipsed by discrete electronics.[96] Under the increasingly apparent grip of these very conditions, labor can paradoxically become what it always has been: the privileged term for the work our robots cannot do, *our* work. As the dialectical antithesis to cybernetic synthesis, this moving target, repeatedly deconstructed and reconstructed by technical innovation over the last 200 years, is vastly more discernable in an environment of technological stagnation. Labor can take its rightful shape as the fulcrum of a global political economy that depends on it in ways both paid and unremunerated across a variety of continents, languages, contexts, and identity positions.

It is in this sense that labor surpasses machines and will always surpass computers as long as they remain bounded by the requirement to function as structurally programmable machines (as we have seen Kittler and Michael Conrad describe in chapter 4). This distinction would hold even were semiconductors to remain on the trajectory of exponential density guiding them for nearly the past sixty years. As such, labor can provide the basis for a new way of organization in a post–Moore's law world. It is that which transcends things as they are, a creative, rather than conservative, force.

It is perhaps for this reason that, under capitalist economies, labor has always been recognized as a threat as much as a resource to manage and harness. For this reason, too, it has a different relationship to entropy than cybernetic systems, even when those systems shackle it to predictable ends. Kittler gets at this, obliquely, when he suggests that "maximizing noise . . . may well be the only way to enter that body of real numbers originally known as chaos."[97] It is also another paradox at the heart of Haraway's endorsement of the cyborg, since she argues that "cyborg politics insist on noise and advocate pollution, rejoicing in the illegitimate fusions of animal and machine."[98] Strictly speaking, however, this is the one thing that cyborgs *cannot* do, at the risk of no longer being cybernetic organisms. The bringing together of disparities in a cybernetic relation does indeed create new and unprecedented fusions, ones that may well appear monstrous or impure (especially from the perspective of more established epistemologies). Yet it is the very act of bringing them into this relation that minimizes noise qua noise and its physical counterpart, entropy, in favor of homeostasis. If we truly want to embrace noise, in other words, we need to enter a world beyond the cybernetic. Labor, from this point of view, is the differentiator between chaos as entropy and chaos as the raw material for an order unobtainable by discrete, electronic, computation alone. In this sense, it stands in opposition to the purely physical principles governing the thermodynamic relation to entropy, even as humans and other animals are ultimately subject to those same laws in strictly physical terms.

To understand labor in this way would represent a striking departure from how postmodern approaches have typically viewed it. Writing as the transformations wrought by the collapse of state socialism removed one of the last vestiges of ideological commitment to the idea of labor from the world stage, Rabinbach argues that the mid-century cybernetic turn effectively undid the earlier discourse of *Arbeitskraft*. "If work is no longer defined by the effects of the 'productive force' of labor power," he writes, "and is no longer subordinated to the goal of the production of things, if the world of signs and communications belongs to human products and productive capacity, work can no longer be understood in terms of material output and performance."[99] In a classically poststructuralist move, however, he argues that this predominance of signs and symbols extends backward to the very metaphors of thermodynamics that proliferated in the nineteenth century, concluding that "to assert that the material reality of the work-centered society has been supplanted by the 'information-' or 'image-centered' society is, however, to ignore the ideological and cultural dimension of the 'materialism' and the centrality of the body in nineteenth-century thought."[100] The persistence of thermodynamics and the foundational concern with optimizing the circulation of energy in cybernetics—and, by extension, computing—goes unremarked here. Moreover, this very framework is ideologized as more of a symbolic effect of Victorian-era discourse than a titanic transformation in the relationship between human beings and the material world brought about by industrialization.

While he may have been premature in diagnosing an end to the concern with work in the postindustrial era, Rabinbach is correct in noting this shift in discursive register. Under circumstances of increasing computational power and neoliberal transformations in industrial relations that have set labor on the back foot, it is perhaps not surprising that a common political thread has centered around the rejection of work and/or labor. Less about the temporary and situational demonstration of the power of labor via the strike than the inhumanity or wastefulness of work entirely, this sentiment can be found in perspectives ranging from the autonomist left to recent reconsiderations of thermodynamics from the perspective of environmental conservation.[101] If rapid increases in the amount of work capable of being performed by microchips over this period have deemphasized the role of labor, conditions of 100% utilization bring to light the difference between these two oft-confused concepts.

As we come to realize what our labor is, we will simultaneously come to terms with the work we delegate to our robots. In redefining accumulation from a private interest to a collective good, a post–100% utilization world would represent an order of political economy above and beyond those predicated upon growth alone. It would not reject machinery, or machinic work, but redefine it as a much-needed tool for the betterment of humanity, one shorn of both its threat and mystique. In this sense, it would revolutionize, rather than render equivalent, relationships between people, technology, and nature. As the inputs of computational work required to sustain it plateau, the time of late-capitalist stagnation is slowly, but surely coming to an end. Beyond homeostasis, another world awaits.

Conclusion: Indeterminate Elements

To what end, Moore's law? The core argument of this book has been that the development of discrete, electronic, silicon computing *has* an end, and more than an end, a trajectory. As the secret teleology behind not only the continuing advancement of capitalism past the point where its manufacturing engine has begun to falter but also many of the theories that have arisen to oppose it, Moore's law represents an unacknowledged, material influence on postindustrial economics and postmodern culture alike. Yet, if computing power has been a hidden determinant of so much of our lives, then a similarly subtle thread of my own argument has been its implications for our understanding of determinism as such. Long a subject of debate, particularly in Marxist circles, the vicissitudes of the concept find an especially revealing home in the *in*determinacy of one of the seemingly most promising developments in post–Moore's law technical infrastructure: quantum computing.

As we saw in the introduction, a transition from silicon to quantum computing does not offer a catch-all solution to the problems of 100% utilization. At the same time, its novelty *does* encourage us to imagine it as a transformative medium, one whose very operation resists any notion that reality is on a set course. The thus far limited capabilities of really existing quantum systems, if anything, make these visions even more revealing than the actual operations of the technology. Let us, return, then, to where we began, for the television show *Devs* is ultimately about much more than the recursive nature of reality in its encounter with computation. It is concerned with nothing less than determinism itself.

A few scenes after the show's quantum simulation of reality comes online, Lily, the series' cryptographer protagonist, describes it as "the tech nerd's wettest dream, the one that reduces everything to nothing, nothing but code. . . . They believe everything is computational. And of course, they've cracked the encryption key to compute it." The system, after which the show is named, is subsequently revealed to have obfuscated its true significance, by way of recourse to the Roman alphabet, as a replacement for God. Devs' totalizing simulation is an extreme example of what Friedrich Kittler, following Michael Conrad, refers to as "the so-called Church-Turing thesis in its strongest or physical form."[1] Perhaps Borges's cartographic recursion is indeed the relevant touchstone here, for, as Conrad explains, "interpreted in its strongest form, the Turing-Church thesis . . . asserts that, as far as presently known, all physically realizable maps can in effect be computed by Turing machines, that is, the systems described by these maps can be simulated to an arbitrarily high degree of precision."[2] Yet, in describing it the way she does, Lily makes a mistake that it is very unlike a cryptographer to make.

In confronting Devs' reconciliation between the world of things and the world of code, one which seemingly transcends the asymptote that Conrad's statement implies, Lily conflates the sense of "code" as procedural instructions with its other commonly understood definition: encoded information. Whatever one may think of the claim that digital encoding reduces information to immateriality, as we have seen throughout this book, it is far less tenable for computation. Cryptographers are usually satisfied if the heat death of the universe is likely to happen before an efficiently programmed computer guesses a key,[3] a factor of the speed of calculation fundamentally determined by the computer's material configuration, its specific existence as a physical thing. Following Turing's acknowledgment of a limit to any analogy between cryptography and physics,[4] however, cracking the code by itself doesn't render computation moot for the same reason that your old computer may still struggle to run a new game it just securely downloaded from an app store.

It is precisely such elisions of computation that encroaching conditions of 100% utilization promise to illuminate. As Conrad puts it, the strong Turing–Church hypothesis is valid only "assuming constraints on space and time bounds are ignored."[5] While it has not been the purpose of this book to adjudicate the question of whether the universe is computable, the shadow of this hypothesis looms large over it. The tendency to equate scientific unknowns with the most sophisticated tools we have to analyze them militates against it.[6] When all you have is the Swiss Army knife of

Turing-completeness, everything looks like a computer science problem. Nevertheless, the astounding capacity that informationalization and computation have shown, over the history of Moore's law, for mediating extensive aspects of reality continues to dangle the possibility tantalizingly before us. The simple reason for leaving the question unanswered is that it is, and perhaps always will be, unfalsifiable. It is precisely this gap between theory and practice that the present volume *has* been concerned with, one that is fully on display in *Devs*.

Limits of time and space are almost entirely absent from the show's view of computation in what is, apart from this, a very scientifically and even philosophically astute narrative. It is important to clarify that this absence is not merely due to the series' invocation of quantum, rather than classical, computing. The capabilities attributed to the Devs system go far beyond the capacity of any computer, quantum or otherwise, as commonly understood. This makes it an exemplary representation of our hopes and fears around determinism as such. Its ability to trace every particle in the universe back to its origin in the Big Bang and forward again, neatly and cryptically alluded to by a neon sign in the company's open-plan, café-like office reading "WE ARE ALL STARDUST," represents a complete triumph over entropy, a reversal of time and an overcoming of the second law of thermodynamics before which even quantum computing's most significant potential achievements cannot but appear modest.[7]

So what *can* quantum computing do? Presently the technology is hampered by the difficulty of maintaining enough quantum bits, or qubits, in a reliable enough state to perform sophisticated calculations, something that engineers are looking to classical, silicon-based computing—and its historically exponential rates of improvement—to help with.[8] Yet it is expected, for example, that quantum computers will render much of the public key encryption we have relied upon over the last thirty years to protect our digital transactions transparent. That is because, with the proper number of qubits, they are able to solve the problems upon which these cryptographic algorithms have been based in polynomial time, problems that were chosen because discrete, electronic, silicon-based computers can only solve them much more slowly, in exponential time. As computer scientist Lance Fortnow makes clear, any improvement from quantum mediation will probably apply only to a subset of the tasks that can currently best be solved by monotonously searching the problem space nondeterministically—that is, heuristically—while *testing* each potential solution in polynomial time. This would leave the fundamental distinction between these two approaches, known as P vs. NP, intact. Quantum computers, he writes, "may help us break codes and help us better understand the fundamental nature of the universe, but they likely won't solve NP-complete problems or make spreadsheets run faster."[9] Even if they will, his description of a world in which P and NP are equivalent (meaning an algorithm capable of solving all NP problems in polynomial time exists and is made known), seems primitive in comparison to Devs. Fortnow sketches this possible, if unlikely, future as one in which the onset of cancer is instantly detected and widely curable, weather and sporting results are largely—though not totally—known in advance, languages are instantly and accurately translatable, lifelike computer simulations proliferate, and many more jobs are automated away than have been already.[10] However, a completely deterministic replica of everything that exists, derived from first principles, dwarfs any solution to this, one of the most important problems in computer science, to which a million-dollar Millennium Prize is attached.

The specter of determinism, that Laplacean night in which all choices are false, haunts the cast of *Devs*. It is depicted as the megalomelancholical obsession of the company's founder, Forest, who seeks, in the perfection of a system proving he had no say in the matter, absolution for his unwitting part in the car-crash death of his family. To most of the other characters, it is an object of sublime horror, a challenge to free will and thus the latest threat to humanity in the age of advanced technical mediation. When Lily's engineer boyfriend exclaims, upon first encountering the system, that "this changes everything," the Devs team lead responds, "It changes absolutely nothing. In a way, that's the point."

It should not surprise that the series' denouement resists this conclusion. First, it is revealed that, much to Forest's chagrin, Devs has only become fully operative by incorporating the "Everett interpretation," quantum physicist Hugh Everett III's theory that the universe splits in two with each measurement or observation. This demands an exponential explosion of possible universes that, while strictly speaking remaining deterministic, in effect raises the prospect of equally possible and even opposite reactions to the same event. What was once a question of causality now becomes one of which timeline one inhabits. Secondly, a space is opened when Lily, just prior to the moment beyond which the system's projections have proven unable to see, makes a slightly different decision than predicted, with minor, though philosophically significant, implications. Choice is thus recuperated and within this system, much like many theorizations of our own late-capitalist nightmare, a sliver of agency and the hope that things didn't *have* to be the way they are can still be held out against the forces of overwhelming totalization.

It is not my intention to simply dismiss *Devs* for this conclusion. As any savvy consumer of fiction, science or otherwise, is aware, creators take great liberties with factual details in service of larger artistic goals. Rather, I want to highlight *which* liberties are taken in service of rescuing the notion of free will. The question of determinism is hardly unrelated to that of a computable universe. As Wendy Hui Kyong Chun has argued, "Computers, understood as software and hardware machines, have made possible a dream of programmability, a return to a world of Laplaceian determinism in which an all-knowing intelligence can comprehend the future by apprehending the past and present."[11] But, as in *Devs*, such a future is thwarted, here by the effects we have already seen of inevitable gaps in translation between one layer and another of the programming hierarchy. "Crucially, though," she writes, "this knowledge is also based on a profound ignorance or ambiguity: our computers execute in unforeseen ways, the future opens to the unexpected. Because of this, any programmed vision will always be inadequate, will always give way to another future."[12] As with a computable universe, and for much the same reasons, I make no claim as to the ultimate status of existence as clocklike or otherwise. Instead, I reserve myself to the much more limited and defensible position, namely that we have a propensity to mistake some determinations as indeterminacy, particularly when doing so would benefit us in some way. This is, in a sense, the opposite, or dialectical counterpart, to Chun's position that we mistake as straightforward the circuitous way in which a high-level programming function like `print()` results, via its traversal of the software–hardware hierarchy, in the appearance of whatever it is to be printed onscreen.[13] Thus, just as she argues that reified source code veils the realities of execution with the intentions of the sovereign programmer, when *Devs* elides the material factors of computation it does so in service of a liberal humanist narrative about the ultimately redemptive power of individual choice—if not for its main characters, who nevertheless end the series fully posthuman, then for its audience.

Another name for this propensity in the Marxist tradition is ideology, a topic I have thus far largely sidestepped in favor of the material determinants of computation. Much could be, and has been, written about this concept, with Chun in particular, as discussed in chapter 4, providing an updated analysis in terms of our misrecognition of the inner workings of computers. In the interest of economy, as it were, I will simply go here to the source: "The phantoms formed in the human brain are also, necessarily, sublimates of their material life-process, which is empirically verifiable and bound to material premises."[14] This basic distinction remains operative even in semiological analyses over a century later. Roland Barthes, for example, in considering the ideological role of myth, offers the example of a newspaper implying that it is not the dynamics of supply and demand, but rather technocratic state intervention that determines produce pricing. "In a first (exclusively linguistic) system," he argues, "causality would be, literally, natural: fruit and vegetable prices fall because they are in season. In the second (mythical) system, causality is artificial, false; but it creeps, so to speak, through the back door of Nature."[15] Notable here is that, within the Hegelian framework in which Marx and Engels are operating, ideological "phantoms" are aligned with abstraction, as is Barthes's *naturalisierungseffekt*. Superficially plausible explanations are attributed in lieu of actual ones, which can only be revealed by a closer analysis of the concrete particulars. In so doing, we may begin to correct this misleading impression. The critique of false consciousness, then, moves from erroneous generalization in service of the status quo back to situational specifics in an effort to ultimately arrive at a more accurate, if arguably no less abstract, conclusion.

Or at least it did. What is striking is the extent to which this schematic has been inverted. Today ideology increasingly wallows in the local minima of the concrete. Take, for example, the late actor–network theorist Bruno Latour, who tweeted that COVID-19 recovery planning should "follow the ways in which the virus . . . spread globally without ever leaving the person to person connection, never shifting to another more 'general' level."[16] In her recent Marxist analysis of contemporary trends that seek to reduce the appearance of mediation and privilege the concrete over the abstract, Anna Kornbluh offers a detailed and incisive criticism of this theoretical tendency: "The flat metaphysics of actor–network theory (ANT), object-oriented ontology, thing theory, posthumanism, and romantic ecological fusionism all directly aim to oust analyses of class relations with maps of distributed agency and to enmesh the knowing subject in hypercomplexity."[17] Everything is too complicated to draw any broad conclusions; there are always exceptions; philosophy is a social graph—this is now the rallying cry of ideology. By sticking to these particulars, we assiduously avoid coming to any overarching determination about the (increasingly deleterious) way that capital conditions flows of money, information, goods, and people.

This fear of big picture thinking connects the repudiation of causality by physicists discussed in chapter 4 (not to mention the postmodern implications those like Lyotard have drawn from it) not only to quantum computing, but the break from Hegelian Marxism pioneered by the French philosopher Louis Althusser. Althusser famously reformulated Marxist dialectics as anti-Hegelian and consisting of, rather than one constitutive opposition, a panoply of contradictions all potentially dispositive at different moments and only determined by the economy "in the last instance."[18] Such an analysis is already primed, as Fredric Jameson pointed out long ago, explicating the contribution of Barry Hindess and Paul Hirst, to disintegrate "into a rubble of autonomous instances without any necessary relationship to each other whatsoever—at which point it follows that one can no longer talk about or draw practical political consequences from any conception of social structure."[19] This is, of course, a desirable effect as far as the entrenched ruling classes that ideology bolsters are concerned. As he observes, however, "anyone who believes" this "is living in an alternative universe."[20]

The appeal to multiple universes is ideological in another way as well. Classically, ideology abstracts away one thing in particular: labor, and its determinant role in social organization, as in Barthes's example of a productive harvest vs. mythical state policy. We might say that *Devs*, for all its insistence on the particularities of the multiverse, excludes something similar, yet as we have seen crucially different: *work*. Specifically, it disavows the work of computation while nonetheless leaning on the trope of computer power. In *Devs'* ideological formulation, in addition to the examples provided earlier, consider finally the computational ramifications of the Everett interpretation. If the universe is constantly splitting, potentially an enormous number of times per second, what was once envisioned as linear determination is now a vast problem space that any computer, quantum or no, seeking to map to our reality must traverse quasi-heuristically. This is a source of extreme consternation for Forest, who summarily fires the engineer responsible for implementing this change. When presented with the system's recreation of Christ on the cross, he angrily counters that there is no guarantee that it is the "real" Jesus—that is, the historical one from our timeline. Even if the simulation differs by only a minute and seemingly inconsequential amount, it nevertheless differs. Given that this interpretation is ultimately the one that obtains within the series' narrative, it is significant that the computational work required to suture some reasonable facsimile of our universe together from an overdetermined, exponentially branching multiverse would be enormously greater, not to mention significantly more tedious, than that of a determinate, causal algorithm.

Althusser's most well-known essay famously redefines ideology as a "hailing," a "practical telecommunication" that both lets its subjects know who they are in its eyes (laborer, professional, clergy, capitalist, etc.) and, in so doing, fashions them as those very subjects.[21] Richard Dienst's deconstructive reading of this theory elaborates: "Ideology requires a short circuit between the singular and the general, so that the *reception of a representation* becomes a sending back—a *representation of a reception*."[22] Again, we encounter here the constitutively false imposition of abstraction upon a concrete particularity that it can never fully name, or, in another register, the incommensurability of language and life, telecommunicated information and material reality. But what about the possibility of the reverse, of the physical world impinging on the infinite play of language? In disavowing the work of computation, *Devs* instead offers a tale of our own agency, determinants be damned.[23]

This is not dissimilar to how things are playing out in our own reality. As computation increases the number and variety of tasks machines can take over from people, we have responded by situating our resistance to this development in an informational register. This has resulted in a modern-day politics based on appealing to monopoly tech platforms to exert stronger data governance. In attempting to delist those we oppose from the global community that computation has connected, ostensibly liberal democracies disavow any entropic tendencies they cannot fully control. Like cybernetics, they seek to displace both class conflict and the threat of fascism with the appearance of perpetually smooth operation. The end of Moore's law threatens this, imperiling the growth in computational capacity that has underwritten the continuing functioning of a system of political economy itself past its prime. The fractiousness of our contemporary politics represents the return of the repressed in more ways than one.

From a basic perspective, everything we know so far about physics tells us it is impossible to halve the size of something like a semiconductor forever. The *telos* of Moore's law will thus come to an end eventually, and as I have tried to show, this terminus appears relatively imminent. But the aspects of teleology that have given it such a negative reputation over the last eighty years are less the notion that a determinate physical process has foreseeable results and more the sense that any teleology completely robs us, individually and collectively, of agency. But it is in fact *capitalism* as totality[24] that, in very real terms, takes this autonomy away from us in its relentless drive to replace labor power with machinic work. In this regard, the onset of 100% utilization represents not an absolute end, but a condition of possibility for an as-yet undetermined future lying beyond the inner workings—fictional or scientific—of our postsilicon computers. As with analyses of prospective environmental and political-economic outcomes from Peter Frase and Adam Greenfield, rather than dictating our fate, there are a number of different directions this technological terminus could take.[25]

To begin with the least desirable, it is certainly conceivable that the delicate balance between fixed and variable capital, and labor and the ownership class, could be upset by the end of Moore's law in a way that does not favor a more socially just outcome. An end to the implicit threat of structural unemployment could be met with increased attempts to violently coerce the working class into selling their labor not simply below its true worth, as capitalism has always done, but below the market rate that, under a stagnant technological regime, would rise astronomically. It could also result in renewed efforts to extend the working day, since digital hardware will be unable to segment it any further. Even if this does not occur, the economic chaos resulting from a breakdown of the capitalist human–machine interface could easily enable the resurgent reactionary movements of the twenty-first century to solidify their power. The collapse of Weimar Germany under the weight of the contradictions of industrial capitalism is, of course, historical evidence of a teleology's failure to unfold in the way it has been foreordained. As noted in chapter 5, this development formed the basis for a reassessment of dialectics in a negative, rather than progressive, light. It was, more importantly, a disaster for humanity that it is imperative to never repeat.

The more hopeful outcome has thus been my guiding light here, since a diminishment of technological determinism, as Jameson's analysis suggests, creates the conditions for meaningful human intervention.[26] In an environment where computing power is completely utilized, the true value of labor, in not only financial but social terms, can be fully recognized. As the quality that both enables and transcends sociotechnical relations, its vitality and necessity become all the more apparent to the extent that computation cannot effect its further replacement. This would represent a dialectical inversion away from the stealthily persistent technological determinism of the Moore's law era and toward a collective organization around ourselves as the motivating factor in socio-economic development. In wresting control of our labors from the grasp of profit-motivated efficiency we reclaim a sense of certainty as to what distinguishes ourselves from our machines. This would approach one of the goals that Theodor Adorno, the foremost proponent of negative dialectics, laid out for philosophy in the post-Auschwitz era, "to bring people's consciousness of themselves to the same state of knowledge that they have of nature, instead of them living like cavemen in thrall to their own knowledge of a cosmos in which the hardly sapient species *homo* makes a helpless go of it."[27] An end to Moore's law provides an opportunity to turn our gaze inwards, secure in the knowledge that our technologies are doing what they need to, and no more.

To do so would represent a synthesis worthy of the term dialectical, with all the connotations of simultaneous conservation and transcendence that its German term, *Aufhebung*, implies. It is in this sense, for example, that the profound, and even potentially anti-Marxist, emphasis on the physical properties of matter that I have relentlessly tried to bring into relation with a more historical-materialist conception can finally turn into its opposite, as "the realization of materialism would mean today the end of materialism, of the blind and degrading dependence of human beings upon material conditions."[28] A post–100% utilization world, with its radically different perspective on both ourselves and our relationship to the physical universe, could indeed dedicate itself to all sorts of conceptual innovations currently unthinkable. In doing so, it would open up a universe of poiesis which includes, but goes far beyond, the increasingly direct engagement with hardware that technical commentators have identified with the end of Moore's law.[29] In its raising of the unique qualities of human labor, it could even do away with the category of the human entirely, something both cybernetic and antihumanist thinkers have tried in vain to do from within the framework of capitalist relations. As Sean Cubitt writes,

> The dominant utopian mechanism today is technology, and its counterfaith is Gaia. It is precisely because that dualism is so potent that it is essential to turn our gaze toward the polity—the assembly of human beings in action—as the site from which might arise any alternative. . . . It is we ourselves who must become other in order to produce an other world. The correlative is that we must cease to be human, and most of all cease to exist as exclusively human polity. . . . The road to that goal, however, must lead through the polis, the humanity of humans, and most of all through our communications in order to imagine a way out of stasis.[30]

A more harmonious association between people, animals, and machines—indeed, between us and nature, our rationality and physical reality—exists beyond the cybernetic relationship rather than within it.

An anticapitalist response to 100% utilization would, then, be exciting precisely to the extent that it has been unshackled from the determinisms of the present mode of production, technological, economic, and social alike. But, of course, there is also a third option, one which is simultaneously and paradoxically both more quotidian and more fantastical. I am speaking, of course, of the maintenance of the status quo. Within the current framework of discrete electronic silicon-integrated circuitry, there may be at least some sound and fury still left. It would be premature to rule future innovation out completely, and especially increased competition for what remains to be gained in performance and market share. At the same time, the situation at press time continues to reflect the increasing salience of an end to Moore's law.

As previously noted, business and leadership woes have exacerbated Intel's loss of the technical edge necessary to remain at the forefront of the dynamic named after its cofounder, which, alongside geopolitical concerns over the location of semiconductor manufacture, has lead to the US government taking a nearly 10% "passive ownership" position in the company.[31] Meanwhile, others try their hand at innovating. Samsung, building on its strengths in solid-state storage, memory, and fabrication, has launched a new attempt to develop its own processor designs after an unsuccessful effort in the 2010s.[32] In the meantime, it is working to raise the profile of its Arm-based Exynos processors. These moves could lessen or potentially end the use of Qualcomm's similarly Arm-based Snapdragon chips in its popular Galaxy smartphones. However, its yields remain behind those of TSMC, and this has spelled trouble not only for Exynos but potentially Samsung's capacity to fabricate Snapdragons as well.[33] This latter development, according to the tech news site *Wccftech*, "indicat[es] that the company's cutting-edge semiconductor division is in complete shambles."[34] The future of any bespoke mobile CPU from the company could therefore also be rocky.

The situation on the GPU side is similar. NVIDIA's 2025 GeForce RTX consumer cards appear to offer increased performance over their 2022–2024 predecessors, but also increased power usage. Like supercomputer maker Cerebras, "the 50-series cards use the same TSMC 4N manufacturing process as the 40-series cards, and increasing the number of transistors without changing the process results in a chip that uses more power."[35] Moore's law is at a standstill here. TSMC and Apple, with access to the fabricator's most advanced techniques—an advantage Samsung, Cerebras, and Qualcomm do not currently have, and which NVIDIA is jockeying for—remain at the cutting edge of miniaturization. At the same time, fierce competition in this space may result in surprising upheavals even as the focus continues moving from raw performance to a balance of performance and efficiency. This is already happening in software. In early 2025, NVIDIA shares briefly went down due to the unexpected release of a freely available neural network developed in China, DeepSeek R1. Trade controls on high-end NVIDIA processors meant its developers were forced to work under constrained computational conditions, and this appears to have resulted in a substantially more efficient piece of software (and cheaper development process) than mainstream Western offerings.[36] This is precisely the kind of innovation that conditions of 100% utilization will ultimately force upon all software. Yet, as we have seen, the fundamental uncertainty of the neural approach encourages near-continual training, so any rush to proclaim the end of the AI hardware market is premature. Thus, while a limit to silicon transistor density improvement is ultimately inevitable, short-term developments remain less predictable.

More speculatively, the scientific community continues to enthusiastically pursue materials that could enable electrical superconductivity under conventional environmental conditions. Such substances, were they to exist, would essentially do away with entropic heat loss in the transfer of electrical energy, with substantial ramifications for electronic circuity, if not semiconductors themselves.[37] They could, in short, realize the cybernetic dream of zero thermodynamic degradation, at the very least substantially boosting the efficiency, and potentially the absolute performance, of discrete, electronic computing devices.[38] This effort, particularly whenever it emerges from its specialized domain and into the realm of popular science, serves as a useful reminder that, whatever the status of Moore's law itself, the struggle between humankind and the second law of thermodynamics is by no means at its end. Likewise, though a blink of an eye when weighed against the entirety of human—let alone physical—history, capitalism has succeeded in establishing itself as a monolithic force in our own time to such an extent that it is always unwise to precipitously rule it out.

The theory of 100% utilization laid out here can nevertheless offer insights about the distinctive qualities of humans, and human labor, that remain valid whichever of these outcomes, or any other, takes shape. Even in an environment where programmable computation takes place many times faster than it does today, the differences between analog and digital connectivity outlined by Kittler and Conrad offer a guide to distinguishing between computational commodities that merely transfer value and the living forces that create it. The cost of simulating one system inside another, understood as a foundational aspect of computation since Turing's formulation, provides another indicator, at once conceptual and physical, of computation's boundaries. So, too, may the limitations to parallel processing represented by Amdahl's law. In a sense, the closer our machines get to approximating us, the more we can be certain of that which separates us from them. This, too, is a dialectic, and one that will eventually end, though none of us can be certain how.

Notes

Introduction

1. His original projection can be found in Gordon E. Moore, "Cramming More Components onto Integrated Circuits" (1965), reprinted in *Proceedings of the IEEE* 86, no. 1 (January 1998): 82–85. It was later updated in Gordon E. Moore, "Progress in Digital Integrated Electronics" (1975), reprinted in *IEEE Solid-State Circuits Society Newsletter* 11, no. 3 (September 2006): 36–37.

2. These issues with heat, clock speed, and density alike take up much of this book, but here see, e.g., M. Mitchell Waldrop, "More than Moore," *Nature* 530 (February 11, 2016): 144–147 and The National Research Council Committee on Sustaining Growth in Computing Performance, *The Future of Computing Performance: Game Over or Next Level?*, eds. Samuel H. Fuller and Lynette I. Millett (Washington, DC: National Academies Press, 2011), 90 (referred to hereafter as *FCP*). Subsequent improvements in size and especially performance have arrived at less than the exponential rates associated with Moore's law in general and twentieth-century integrated circuits specifically (see, e.g., Andrew Cunningham, "Intel Tries to Get Its Chip Manufacturing Back on Track with 'Intel 4,' Due in 2023," *Ars Technica* June 13, 2022, https://arstechnica.com/gadgets/2022/06/intel-tries-to-get-its-chip-manufacturing-back-on-track-with-intel-4-due-in-2023/, accessed June 14, 2022). Strictly speaking, however, Moore's law refers only to semiconductor size, not performance.

3. John Shalf quoted in Waldrop, "More than Moore," 146.

4. In theory, it is the role of programmers to develop procedures that do not needlessly consume resources (see note 71 in chapter 3). In practice, it is often systems administrators—a previous profession of mine that informs my analysis here—who are tasked with ameliorating the effects of an out-of-control program monopolizing processor time.

5. Karl Marx, *Grundrisse: Foundations of the Critique of Political Economy (Rough Draft)* (1857–1858) trans. Martin Nicolaus (1973) (London: Penguin, 1993), 690–712. For a speculative treatment of these themes, however, see the fascinating short story

by Chinese science-fiction writer Cixin Liu, "For the Benefit of Mankind" (2005) in *The Wandering Earth* (2013), trans. Ken Liu, Elizabeth Hanlon, Zac Haluza, Adam Lanphier, and Holger Nahm (London: Head of Zeus, 2017), 155–208. (The video game *The Last Worker* [2023] explores similar territory.) Liu's most well-known work hinges on a computationally intractable phenomenon (Cixin Liu, *The Three-Body Problem* (2006), trans. Ken Liu (New York: Tor, 2014)).

6. Karl Marx, *Capital: A Critique of Political Economy*, vol. 3 (1863–1883/1894) trans. David Fernbach (1981) (London: Penguin, 1991), 315–375. Both theorizations and their modern-day adherents are discussed in chapter 5.

7. As media theorist Friedrich Kittler suggested in a lecture toward the end of his life, Moore's law, "is bound to fail one day" following "a point that Thomas Pynchon's novels make and make again and again, that . . . an economy of exchange . . . is an exponential one, that is [a] capitalistic economy has a built-in breakpoint, failure point. It's asymmetrical from the start and the dreams one day have to disappear" (Friedrich Kittler, "Computing, Ubiquity and the One Single Medium" 2010 lecture at the European Graduate School, https://www.youtube.com/watch?v=WYn-FKRA0mn8). Such statements make clear the extent to which critiques of Kittler as apolitical or right-wing miss the mark.

8. E.g., Paul Cockshott, Lewis M. Mackenzie, and Greg Michaelson, *Computation and Its Limits* (Oxford: Oxford University Press, 2012). For their treatment of the hardware issues described above, written before some of the latest technological developments (e.g., FinFET devices enabling reduced gate capacitance), see 2–3, 90–94, and 116–117.

9. Ron Eglash, "Computing Power," in *Software Studies: A Lexicon*, ed. Matthew Fuller (Cambridge, MA: MIT Press, 2008), 61.

10. E.g., Jussi Parikka, *A Geology of Media* (Minneapolis: University of Minnesota Press, 2015); Sean Cubitt, *Finite Media: Environmental Implications of Digital Technologies* (Durham, NC: Duke University Press, 2017).

11. See, e.g., the essays in Lisa Parks and Nicole Starosielski, eds., *Signal Traffic: Critical Studies of Media Infrastructures* (Urbana, IL: University of Illinois Press, 2015), as well as the two monographs by Starosielski, *The Undersea Network* (Durham, NC: Duke University Press, 2015) and *Media Hot & Cold* (Durham, NC: Duke University Press, 2021). The type of symptomatic, structuring absence outlined here is traceable back to Louis Althusser's reinterpretation of Marx in Louis Althusser, Etienne Balibar, Roger Establet, Pierre Macherey, and Jacques Rancière, *Reading* Capital (1965) Complete Edition (1996) trans. Ben Brewster and David Fernbach (London: Verso, 2016), e.g., 22–53.

12. Matthew Kirschenbaum, *Mechanisms: New Media and the Forensic Imagination* (Cambridge, MA: MIT Press, 2008).

13. E.g., M. Beatrice Fazi, *Contingent Computation: Abstraction, Experience, and Indeterminacy in Computational Aesthetics* (London: Rowman and Littlefield, 2018); Alexander R. Galloway, *Uncomputable: Play and Politics in the Digital Age* (London: Verso, 2021).

14. E.g., the MIT Press's excellent Platform Studies series, in particular books such as Nick Montfort and Ian Bogost, *Racing the Beam: The Atari Computer System*; Jimmy Maher, *The Future Was Here: The Commodore Amiga* (both 2012); and Nathan Altice, *I Am Error: The Nintendo Family Computer/Entertainment System Platform* (Cambridge, MA, 2015). A similar sensibility seems to be motivating reconsiderations of the burgeoning independent and/or illegal telecommunications scenes of the 1980s and early 1990s, including Kevin Driscoll, *The Modem World: A Prehistory of Social Media* (New Haven, CT: Yale University Press, 2022) and Martin Paul Eve, *Warez: The Infrastructure and Aesthetics of Piracy* (Goleta, CA: Punctum Books, 2021).

15. E.g., Florian Cramer, "What Is 'Post-Digital'?" *APRJA* 3, no. 1 (2014), https://aprja.net/article/view/116068/165295, accessed April 25, 2020; Lauren Berlant, *Cruel Optimism* (Durham, NC: Duke University Press, 2011); Wendy Hui Kyong Chun, *Updating to Remain the Same: Habitual New Media* (Cambridge, MA: 2016). In a more explicitly Marxist vein, see Mark Fisher, *Capitalist Realism: Is There No Alternative?* (Winchester, UK: Zero Books, 2009) and Simon Reynolds, *Retromania: Pop Culture's Addiction to Its Own Past* (New York: Faber and Faber, 2011).

16. E.g., Wolfgang Streeck, *Buying Time: The Delayed Crisis of Democratic Capitalism* (2013) 2nd ed., trans. Patrick Camiller and David Fernbach (London: Verso, 2017) and Thomas Piketty, *Time for Socialism: Dispatches from a World on Fire, 2016–2021*, trans. Kristin Couper (New Haven, CT: Yale University Press, 2021).

17. Karl Marx, *Capital: A Critique of Political Economy*, vol. 1 (1867), trans. Ben Fowkes (1976) (London: Penguin, 1990), 164.

18. E.g., Maurizio Lazzarato, "Immaterial Labor," in *Radical Thought in Italy: A Potential Politics*, trans. Paul Colilli and Ed Emory, ed. Paolo Virno and Michael Hardt (Minneapolis: University of Minnesota Press, 1996), 133–147. Seb Franklin argues that a cybernetic and "information-theoretical configuration . . . exemplifies the fantasy that all forms of labor can be productively imagined as idealized processes of selection from a set of discrete symbols. The conflation of information and work posits all labor as immaterial labor in the sense that Lazzarato defines it" (Seb Franklin, "Cleverness and Drive, or the Cybernetic Fantasy of Value: R.S. Hunt's 'Two Kinds of Work,'" *Grey Room* 68 (Summer 2017): 48 (discussed further in chapter 6). I am in a sense making the inverse claim, that "immaterial" labor is simply one subset of labor in general.

19. Giovanni Arrighi, *The Long Twentieth Century: Money, Power and the Origins of Our Times* (1994) New and Updated Edition (London: Verso, 2010), 6. See also Marx, *Capital*, vol. 1, 246–257.

20. Arrighi, *The Long Twentieth Century*, 9.

21. Fredric Jameson, "Culture and Finance Capital" in *The Cultural Turn: Selected Writings on the Postmodern, 1993–1998* (London: Verso, 1998), 143.

22. Ibid., 143–144.

23. Including by Jameson in an older text: Fredric Jameson, *The Political Unconscious: Narrative as a Socially Symbolic Act* (Ithaca, NY: Cornell University Press, 1981), 45–46.

24. Marx, *Capital*, vol. 1, 173; Jane Bennett, *Vibrant Matter: A Political Ecology of Things* (Durham, NC: Duke University Press, 2010) and Graham Harman, *Object-Oriented Ontology: A New Theory of Everything* (London: Pelican, 2018).

25. See Steven Shaviro, "Consequences of Panpsychism," in *The Nonhuman Turn*, ed. Richard Grusin (Minneapolis: University of Minnesota Press, 2015), 19–20 and also, more critically, note 17 in the conclusion.

26. Anne Pasek, "From Atoms to Electrons: An Energy History and Future of Computing," in *Digital Energetics*, ed. Anne Pasek et al. (Lüneburg, Germany/Minneapolis: meson/University of Minnesota Press, 2023), 26–42. Her chapter offers an alternative, but broadly compatible, analysis of Moore's law and many of the processes discussed here from the perspective of energy as a medium and its centrality to computation. For another analysis of when the ultimate limits to microchip energy efficiency and speed could be reached (predicting roughly 2045 and 2050, respectively), see Cockshott et al., *Computation and Its Limits*, 116–117.

27. See, e.g., Cara New Daggett, *The Birth of Energy: Fossil Fuels, Thermodynamics, and the Politics of Work* (Durham, NC: Duke University Press, 2019) and Kohei Saito, *Marx in the Anthropocene: Towards the Idea of Degrowth Communism* (Cambridge, UK: Cambridge University Press, 2022).

28. Ståle Holgersen, "Neither Productivism nor Degrowth: Thoughts on Eco-socialism," *Spectre* September 4, 2023, https://spectrejournal.com/neither-productivism-nor-degrowth/, accessed August 13, 2024.

29. Jorge Luis Borges with Adolfo Bioy Casares, "Of Exactitude in Science" (1946), in *A Universal History of Infamy*, trans. Norman Thomas di Giovanni (1954) (New York: E. P. Dutton and Co., 1972), 141; Jean Baudrillard, *Simulacra and Simulation* (1981), trans. Sheila Faria Glaser (Ann Arbor, MI: University of Michigan Press, 1994).

30. Jean-François Lyotard, *The Postmodern Condition: A Report on Knowledge* (1979), trans. Geoff Bennington and Brian Massumi (Minneapolis: University of Minnesota Press, 1984), 55.

31. Ibid., 67.

32. For a contemporary treatment of this history see Tung-Hui Hu, *A Prehistory of the Cloud* (Cambridge, MA: MIT Press, 2015), 37–71. Tellingly for the present study,

time-sharing systems, which allow multiple users to interact with the same (stand-alone) server independently, were developed out of a sense that computers were becoming so much faster than humans that the former could reliably juggle between the latter (see Fernando J. Corbató, Marjorie Merwin-Daggett, and Robert C. Daley, "An Experimental Time-Sharing System," *Proceedings of the American Institute of Electrical Engineers–Institute of Radio Engineers Spring Joint Computer Conference* (May 1–3, 1962): 335–344).

33. E.g., Tero Karppi, *Disconnect: Facebook's Affective Bonds* (Minneapolis: University of Minnesota Press, 2018); Safiya Umoja Noble, *Algorithms of Oppression: How Search Engines Reinforce Racism* (New York: NYU Press, 2018).

34. A similar issue, discussed further in chapter 4, constrains many critiques of artificial intelligence.

35. E.g., Benjamin H. Bratton, *The Stack: On Software and Sovereignty* (Cambridge, MA: MIT Press, 2015) and Jonathan Beller, *The World Computer: Derivative Conditions of Racial Capitalism* (Durham, NC: Duke University Press, 2021). Bratton admits that "the limits to growth for The Stack" include "Moore's law" (Bratton, *The Stack*, 92).

36. Alexander R. Galloway, *The Interface Effect* (Cambridge, UK: Polity, 2012), 8.

37. Lev Manovich, *The Language of New Media* (Cambridge, MA: MIT Press, 2001). For a more recent consideration see Jonathan Sterne, *MP3: The Meaning of a Format* (Durham, NC: Duke University Press, 2012), 194–196.

38. Friedrich Kittler, "There Is No Software," *Stanford Literature Review* 9.1 (spring 1992): 81–90; Wendy Hui Kyong Chun, *Programmed Visions: Software and Memory* (Cambridge, MA: MIT Press, 2010).

39. E.g., Warren Sack, *The Software Arts* (Cambridge, MA: MIT Press, 2019); Justin Joque, *Deconstruction Machines: Writing in the Age of Cyberwar* (Minneapolis: University of Minnesota Press, 2018); and Winnie Soon and Geoff Cox, *Aesthetic Programming: A Handbook of Software Studies* (London: Open Humanities Press, 2020), among numerous others. For a similar, but broader, critique, of the emphasis on symbolic over physical factors in digital media, including how both aspects are dependent on unsupportable growth, see Cubitt, *Finite Media*, 85–6.

40. Sterne, *MP3*, 201.

41. Alexander R. Galloway, *Protocol: How Control Exists After Decentralization* (Cambridge, MA: MIT Press, 2004), 165–166; Chun, *Programmed Visions*, 23 (see also the previous and following pages). Chun is also responding to assertions made in Alexander R. Galloway, "Language Wants to Be Overlooked: On Software and Ideology," *Journal of Visual Culture* 5, no. 3 (December 2006), 315–331, published in revised form in Galloway, *The Interface Effect*, 54–77. Kittler's original statement, that "there exists no word in any ordinary language which does what it says" implies Galloway's conclusion but instead directly and deconstructively refers to the necessary

inaccuracy of Kittler's essay, a natural-language attempt to consider "hardware" in terms of the university's role in knowledge formation (Friedrich Kittler, "On the Implementation of Knowledge—Toward a Theory of Hardware," nettime mailing list post, date unknown, archived at http://hydra.humanities.uci.edu/kittler/implement.html, accessed June 14, 2022).

42. Chun, *Programmed Visions*, 25, 26.

43. Eglash, "Computing Power," 57–58.

44. This avoids any speedups from cached machine instructions.

45. "Moore's law helped to drive the progression in sequential-language abstractions because increasing processor speed hid their costs" (*FCP*, 106).

46. Nick Bostrom, "Are We Living in a Computer Simulation?" *The Philosophical Quarterly* 53, no. 211 (April 2003): 253.

47. Ibid.

48. See Michael Conrad, "The Price of Programmability," in *The Universal Turing Machine: A Half-Century Survey*, ed. Rolf Herken (1988), 2nd ed. (Vienna: Springer-Verlag, 1994/1995), 261–281. A reading of Kittler, "There Is No Software" and Conrad's influence upon it thus forms a substantial part of chapter 1 and returns, focusing on a different aspect, in chapter 4.

49. Philip Rosen, *Change Mummified: Cinema, Historicity, Theory* (Minneapolis: University of Minnesota Press, 2000), 329–330.

50. Ibid., 348. See also Chun, *Programmed Visions*, 50–54.

51. Rosen, *Change Mummified*, 329.

52. My first chapter thus considers recent, sober analyses of this problem from the computer science and electrical engineering community (but for more historical acknowledgments see, e.g., Donald E. Knuth, *Things a Computer Scientist Rarely Talks About* (1999/2001) (Stanford, CA: Center for the Study of Language and Information, 2003), 16.). Chapter 5, by contrast, is more directly concerned with ideologies of computing power found in and derived from Silicon Valley.

Chapter 1

1. Ernest Mandel, "Introduction," to Karl Marx, *Capital: A Critique of Political Economy*, vol. 1 (1867) trans. Ben Fowkes (1976) (London: Penguin, 1990), 60.

2. Marx, *Capital* vol. 1, 784.

3. Fredric Jameson, *Representing* Capital*: A Reading of Volume One* (London: Verso, 2011), 148. On Marx's reserve army and the decaying welfare state see also Aaron Benanav, *Automation and the Future of Work* (London: Verso, 2019), 47–48.

4. See, for example, Olivia Solon, "Self-Driving Trucks: What's the Future for America's 3.5 Million Truckers?" *The Guardian*, June 17, 2016, https://www.theguardian.com/technology/2016/jun/17/self-driving-trucks-impact-on-drivers-jobs-us, accessed March 30, 2017 and Adam Greenfield, *Radical Technologies: The Design of Everyday Life* (London: Verso, 2017), 192–193. Uber and Alphabet subsidiary Waymo, which sued Uber for stealing trade secrets with its acquisition of Otto, a company founded by a former Google autonomous vehicle engineer, continue to pursue self-driving technologies.

5. Sam Elitzer quoted in Solon, "Self-Driving Trucks."

6. The canonical text here is Karl Marx, "Preface to *A Contribution to the Critique of Political Economy*" (1859) in *Early Writings*, trans. Rodney Livingstone and Gregor Benton (London: Penguin, 1992), 424–428. On the crucial significance of this dialectic to Marxism, see Fredric Jameson, "Three Names of the Dialectic," in *Valences of the Dialectic* (London: Verso, 2009), 43–48 and Jameson, *Representing* Capital.

7. Randall Collins, "The End of Middle-Class Work: No More Escapes," in *Does Capitalism Have a Future?* ed. Immanuel Wallerstein et al. (Oxford: Oxford University Press, 2013), 41. In his more popular account of automation, Peter Frase similarly argues that "the labor required to create a pattern that can be infinitely replicated is orders of magnitude less than the labor required in a physical production process in which the same object is made over and over again" (Peter Frase, *Four Futures: Visions of the World after Capitalism* (London: Verso, 2016), 84).

8. Collins, "The End of Middle-Class Work," 38, 57. On AI, specifically, see 41 and 67. See also Greenfield, *Radical Technologies*, 184–185.

9. See, e.g., Arwa Mahdawi, "Will AI Free Us from Drudgery—or Leave Us Jobless and Hungry?" *The Guardian*, May 30, 2023, https://www.theguardian.com/commentisfree/2023/may/30/will-ai-free-us-from-drudgery-or-leave-us-jobless-and-hungry, accessed June 4, 2023.

10. The benefits of this continual refinement for programmers and, by extension, end users, at least up until about seventeen years ago, have been described by Herb Sutter as a "free lunch" (Herb Sutter, "The Free Lunch Is Over: A Fundamental Turn Toward Concurrency in Software" (December 2004/2009), http://www.gotw.ca/publications/concurrency-ddj.htm, accessed April 3, 2017, also published in *Dr. Dobb's Journal* 30, no. 3 (March 2005): 16–22). See also the report by The National Research Council Committee on Sustaining Growth in Computing Performance, *The Future of Computing Performance: Game Over or Next Level?*, ed. Samuel H. Fuller and Lynette I. Millett (Washington, DC: National Academies Press, 2011), 106–109 (referred to hereafter as *FCP*); Paul E. Ceruzzi, *Computing: A Concise History* (Cambridge, MA: MIT Press, 2012), 64; and M. Mitchell Waldrop, "More than Moore," *Nature* 530 (February 11 2016): 145.

11. In making this argument I am inspired by, but also move beyond, musicologist Jacques Attali's analogization of musical performance, as labor, to computational execution (see Jacques Attali, *Noise: The Political Economy of Music* (1977) trans. Brian Massumi (Minneapolis: University of Minnesota Press, 1985), 66, and Andrew Lison, "New Media, 1989: Cubase and the New Temporal Order," *Computational Culture* 8 (July 2021), http://computationalculture.net/new-media-1989-cubase-and-the-new-temporal-order/).

12. *FCP*, 133. Similarly, if more domestically, "The implications of a dramatic slowdown in how quickly computer performance is increasing—for our economy, our military, our research institutions, and our way of life—are substantial" (*vii*).

13. Ibid., 78.

14. Gordon E. Moore, "Cramming More Components onto Integrated Circuits" (1965), reprinted in *Proceedings of the IEEE* 86, no. 1 (January 1998): 83. Moore's article is also included as an appendix to *FCP*.

15. Moore, "Cramming More Components onto Integrated Circuits," 83.

16. Ibid.; Gordon E. Moore, "Progress in Digital Integrated Electronics" (1975), reprinted in *IEEE Solid-State Circuits Society Newsletter* 11, no. 3 (September 2006): 37.

17. Moore, "Cramming More Components onto Integrated Circuits," 83. David C. Brock notes, however, drawing upon Christophe Lécuyer, *Making Silicon Valley: Innovation and the Growth of High Tech, 1930–1970* (Cambridge, MA: MIT Press, 2006) and interviews with Moore, that integrated circuits, in automating previous manufacturing methods, "encompassed the costs associated with a set of the equivalent discrete components and the labor to interconnect them" (David C. Brock, "A Clear Voice: The Origins of Gordon Moore's 1965 Paper" in *Understanding Moore's Law: Four Decades of Innovation*, ed. David C. Brock (Philadelphia: Chemical Heritage Press, 2006), 29 and 36n11. See also his discussion of C. Harry Knowles on 31–34. This volume also includes reprints of the two aforementioned pieces by Moore, as well as the first draft of his 1965 article and additional commentary).

18. Gordon E. Moore, "The Future of Integrated Electronics" (1965), manuscript draft reprinted in Brock ed., *Understanding Moore's Law*, 37 (see also 50 and 52; only the latter reference made it into "Cramming More Components onto Integrated Circuits," 84). Moore here, at least initially, would seem to have been in agreement with Marx. As Jameson notes, "in *Capital* the word 'revolution' always means a technological revolution in the introduction of new and more productive and destructive kinds of machinery" (Jameson, *Representing* Capital, 38). See also Jameson, "Three Names of the Dialectic," 46.

19. Moore, "Cramming More Components onto Integrated Circuits," 82.

20. Gordon E. Moore, "Moore's Law at 40" in Brock ed., *Understanding Moore's Law*, 81. Likewise, "Faster computers create not just the ability to do old things faster but the ability to do new things that were not feasible at all before" (*FCP*, 25).

21. Brock, "A Clear Voice," 28; see also the previous page; Douglas C. Engelbart, "Microelectronics and the Art of Similitude," *1960 International Solid-State Circuits Conference Digest of Technical Papers* (New York: Lewis Winner, 1960), 76–77; and John Markoff, "It's Moore's Law, but Another Had the Idea First," *New York Times*, April 18, 2005, https://www.nytimes.com/2005/04/18/technology/its-moores-law-but-another-had-the-idea-first.html, accessed February 28, 2017 (both texts are cited in Brock, "A Clear Voice").

22. Engelbart, "Microelectronics and the Art of Similitude," 76, 77 (see also Brock, "A Clear Voice," 28); G. W. F. Hegel, *Phenomenology of Spirit* (1807) trans. A. V. Miller (Oxford: Oxford University Press, 1979).

23. *FCP*, 65–66.

24. As Markoff writes, Engelbart's "paper convinced him that the Augmentation machine he envisioned would be possible, because computing would be plentiful in the future" (Markoff, "It's Moore's Law, but Another Had the Idea First").

25. See Jean-François Lyotard, *The Postmodern Condition: A Report on Knowledge* (1979), trans. Geoff Bennington and Brian Massumi (Minneapolis: University of Minnesota Press, 1984).

26. Ceruzzi, *Computing*, 99. Echoing Jameson's considerations discussed in note 6 earlier, Ceruzzi also offers an assessment of the contradictory forces of technological and social determinism specifically vis-à-vis Moore's law: "The historical record suggests that both are correct, a paradox as puzzling as Zeno's but even harder to disentangle" (*xv*; see also the previous page and 111–112).

27. Probir K. Bondyopadhyay, "Moore's Law Governs the Silicon Revolution," *Proceedings of the IEEE* 86, no. 1 (January 1998): 78 and figure 1; see also Moore, "Moore's Law at 40," 67. The preface to Brock ed., *Understanding Moore's Law* argues that "with the proliferation of silicon chips into nearly every aspect of contemporary life, Moore's law serves as an emblem for the whole of technological change" (*ix*).

28. On this latter point, see Lison, "New Media, 1989."

29. Friedrich Kittler, "There Is No Software," *Stanford Literature Review* 9.1 (Spring 1992): 81–90. See also A. M. Turing, "On Computable Numbers, with an Application to the Entscheidungsproblem" (1936), *Proceedings of the London Mathematical Society* ser. 2, vol. 42 (1937): 230–265.

30. Kittler, "There Is No Software," 81.

31. Ibid., 82. The actual development of—and credit for—the 4004 is more complicated than Kittler's narrative suggests (see, e.g., Andrew Lison, "Hardware Standardization and State-Socialist Piracy: The Global Reach of the Zilog Z80," *IEEE Annals of the History of Computing* 46, no. 2 (April–June 2024): 40–41, 47–48). Similarly, the replacement of hand-drafting with computer-aided IC design did not take place immediately after the 4004's arrival. For Marxist computer scientist Paul Cockshott, however, the Turing machine already itself represents a material leap: "Turing's 1936 paper on the computable real numbers marks the epistemological break between idealism and a materialism in mathematics. . . . Turing starts a philosophical tradition of grounding mathematics on the material and hence ultimately on what can be allowed by the laws of physics" (Paul Cockshott, "Turing: The Irruption of Materialism into Thought," *Oxford University Press* blog post June 19, 2012, https://blog.oup.com/2012/06/turing-the-irruption-of-materialism-into-thought/, accessed May 10, 2017). Alain Badiou has argued, however, that the materiality of the Turing machine is solely possible via the ideation of its working space as unlimited (Alain Badiou, "Infinitesimal Subversion" (1968) trans. Robin Mackay with Ray Brassier in *Concept and Form: Key Texts from the* Cahiers pour l'Analyse vol. 1, eds. Peter Hallward and Knox Peden (London: Verso, 2012), 187). Andrew Goffey perhaps synthesizes these positions when he notes that "algorithms bear a crucial, if problematic, relationship to material reality . . . tacit in the way that the Turing machine was envisaged in terms of effective processes: A computer is a machine, after all, and while the Turing machine is an imaginative abstraction, its connotations of materiality are entirely real" (Andrew Goffey, "Algorithm," in *Software Studies: A Lexicon*, ed. Matthew Fuller (Cambridge, MA: MIT Press, 2008), 16). See also Kittler, "There Is No Software," 87.

32. Ibid., 88.

33. Ibid., 89, citing Conrad, "The Price of Programmability" in *The Universal Turing Machine: A Half-Century Survey* (1988), ed. Rolf Herken, 2nd ed. (Vienna: Springer-Verlag, 1994/1995), 268, 265 (these page numbers differ from Kittler, who, referencing the first edition of this text, cites pp. 293 and 290). Kittler's text, which seems to have been developed from spoken remarks, repeatedly renders "price" as "prize," although the former meaning is clear from the context.

34. Conrad, "The Price of Programmability," 264.

35. Ibid. See also Kittler, "There Is No Software," 88–89. More specifically, *"A real system is (effectively) programmable if it is possible to communicate desired programs to it in an exact manner (without approximation) using a finite set of primitive operations and symbols"*; *"a physical system is structurally programmable if it is effectively programmable and if its program is mapped by its structure."* Kittler builds upon Conrad's argument "that it is usually necessary to pay for this decomposability and universality in terms of the potential efficiency with which the material resources of the system are utilized" (Conrad, "The Price of Programmability," 263, 264, 265, emphases in original).

36. Kittler, "There Is No Software," 89, citing Conrad, "The Price of Programmability," 279 (304 in the first edition) on this last point. W. Daniel Hillis emphasizes computing's reliance on "*functional abstraction*. . . . Computers are understandable because you can focus on what is happening at one level of the hierarchy without worrying about the details of what goes on at the lower levels" (W. Daniel Hillis, *The Pattern on the Stone: The Simple Ideas That Make Computers Work* (1998), 2nd ed. (New York: Basic Books, 2015), *xiii*). See also *FCP*, 106–109.

37. Kittler, "There Is No Software," 90. On the sources of noise in integrated circuits he cites Conrad, "The Price of Programmability," 278 (pp. 303–304 in the first edition).

38. Here see Hillis, *The Pattern on the Stone*, 65–66. He argues that, semantically, "any analog signal is bound to be affected by a certain amount of noise . . . while there are an infinite number of possible signal levels, only a finite number of levels represent meaningful distinctions" (65). Kittler's point, however, is a materialist one, that this "noise" *matters*, in both senses of the word.

39. Kittler, "There Is No Software," 82.

40. See Mark Bohr, "14 nm Process Technology: Opening New Horizons," 2014 Intel Developer Forum Presentation, available online at https://www.intel.com/content/dam/www/public/us/en/documents/technology-briefs/bohr-14nm-idf-2014-brief.pdf and Ars Staff, "Intel Is Keeping Moore's Law Alive by Making Bigger Improvements Less Often," *Ars Technica*, March 29, 2017, https://arstechnica.com/information-technology/2017/03/intel-is-keeping-moores-law-alive-by-making-bigger-improvements-less-often/ (URLs accessed July 27, 2025).

41. Usman Pirzada, "Intel 10nm Chip Core i3-8121U Sisoft Sandra Benchmarks Leaked From Lenovo Labs," *Wccftech*, May 16, 2018, https://wccftech.com/intel-10-nm-benchmarks-leak-8121-u/; Joel Hruska, "At Least One of the Recent Intel Roadmap Rumors Is Likely Wrong," *ExtremeTech*, April 10, 2020, https://www.extremetech.com/computing/309129-at-least-one-of-the-recent-intel-roadmap-rumors-is-likely-wrong (URLs accessed April 26, 2020); Jim Salter, "As 7nm Schedule Continues Slipping, Intel Contemplates 3rd-party Fabs," *Ars Technica*, July 24, 2020, https://arstechnica.com/gadgets/2020/07/as-7nm-schedule-continues-slipping-intel-contemplates-3rd-party-fabs/, accessed August 2, 2020.

42. Intel, "Intel 4 Versus Intel 7 and Intel Process Technology Naming," https://www.intel.com/content/www/us/en/support/articles/000097598/processors.html accessed July 19, 2024.

43. Ibid.; Areej, "Intel Renames 10nm ESF Node to 7nm, 7nm to 4nm, 5nm to 3nm in Roadmap Update," *Hardware Times*, July 27, 2021, archived at https://web.archive.org/web/20210726230314/https://www.hardwaretimes.com/intel-renames-10nm-esf-node-to-7nm-7nm-to-4nm-5nm-to-3nm-in-roadmap-update/, accessed July 27, 2025.

44. *FCP*, for example, "believes that the slowing in the exponential growth in computing performance, while posing great risk, may also create a tremendous opportunity for innovation in diverse hardware and software infrastructures that excel as measured by other characteristics, such as low power consumption and delivery of throughput cycles" (54; see also box 2.1 (56–57)). On mobile chips see also Waldrop, "More than Moore," 147.

45. Intel Corporation, "Intel Tick-Tock Model," archived at https://web.archive.org/web/20170528092942/http://www.intel.com/content/www/us/en/silicon-innovations/intel-tick-tock-model-general.html; Intel Corporation, 2015 Annual Report, 12, https://www.intc.com/filings-reports/annual-reports/content/0000050863-15-000015/0000050863-15-000015.pdf (URLs accessed July 27, 2025). For background on and analysis of this development see Brett Howse and Ryan Smith, "Tick Tock on the Rocks: Intel Delays 10nm, Adds 3rd Gen 14nm Core Product 'Kaby Lake,'" *Anandtech*, July 16, 2015, http://www.anandtech.com/show/9447/intel-10nm-and-kaby-lake; Ashraf Eassa, "Intel Corp. Officially Kills 'Tick-Tock,'" *The Motley Fool*, March 22, 2016, https://www.fool.com/investing/general/2016/03/22/intel-corp-officially-kills-tick-tock.aspx; Ian Cutress, "Intel's 'Tick-Tock' Seemingly Dead, Becomes 'Process-Architecture-Optimization,'" *Anandtech*, March 22, 2016, http://www.anandtech.com/show/10183/intels-tick-tock-seemingly-dead-becomes-process-architecture-optimization (URLs accessed May 10, 2017); and Staff, "Intel Is Keeping Moore's Law Alive."

46. Intel Corporation, 2015 Annual Report, 12. A recurring theme of the NRC report, which even includes in its assessment of technological progress the possibility of self-driving military vehicles (*FCP*, 39), is the thoroughgoing significance of the digital turn. For example: "Without performance incentives to buy the next generation of hardware, the economic virtuous cycle is likely to break down, and this would have widespread negative consequences for many industries" (78; see also Waldrop, "More than Moore," 147) and "The impact of computer technology is so widespread that it is nearly impossible to overstate its importance" (*FCP*, 25). On this point see also Brock's summary of then-Intel executive Patrick Gelsinger's remarks at a 2005 Moore's law 40th anniversary conference (David C. Brock, "Reflections on Moore's Law," in *Understanding Moore's Law*, ed. Brock, 101–102) and Hillis, *The Pattern on the Stone*, xiv.

47. Waldrop, "More than Moore," 145. The report was released later that year (Semiconductor Industry Association, *International Technology Roadmap for Semiconductors 2.0: 2015 Edition* Executive Report, https://www.semiconductors.org/resources/2015-international-technology-roadmap-for-semiconductors-itrs/, accessed April 25, 2020). The NRC report argues that "exponential growth of the type and scale that have fueled the entire information technology industry is ending" (*FCP*, 53).

48. Waldrop, "More than Moore," 145. Updating some of his concluding statements from "There Is No Software" quoted earlier, Kittler makes a similar observation (Friedrich Kittler, "Computing, Ubiquity and the One Single Medium," 2010

lecture at the European Graduate School, https://www.youtube.com/watch?v=WYnFKRA0mn8, accessed June 19, 2020). As we have seen, the conclusion to the former essay also offers dissipating heat, or thermodynamic entropy, as a factor (see also Waldrop, "More than Moore," 145, 146, 147 and *FCP*, 80–81). Quantum issues also plagued the move from vacuum to solid-state switches (see Ceruzzi, *Computing*, 66 and 68 as well as David C. Brock, "The Backdrop to Moore's Law: Developments in Semiconductor Electronics to 1965," in *Understanding Moore's Law*, ed. Brock, 8). They have presented a problem for more recent attempts to perpetuate Moore's law as well (Moore, "Moore's Law at 40," 82–84). Efforts to take advantage of quantum tunneling are underway, although preliminary (*FCP*, 98–99; see also Conrad, "The Price of Programmability," 278).

49. Waldrop, "More than Moore," 145.

50. These possibilities are all described in ibid., 146.

51. Waldrop, "More than Moore," 145. See also *FCP*, 98–100 and the summarization of Gelsinger's remarks in Brock, "Reflections on Moore's Law," 102 as well as *FCP*, 67. In the present volume, chapter 4 supplies a deeper examination of attempts to approximate neurology in silicon, while the conclusion will offer a critique of quantum computing.

52. Mark Pesce, "Zombie Moore's Law Shows Hardware Is Eating Software," *The Register*, September 22, 2016, https://www.theregister.com/2016/09/22/the_evolution_of_moores_law_suggests_hardware_is_eating_software/, accessed July 27, 2025. Waldrop's conclusion is similarly hopeful (Waldrop, "More than Moore," 147).

53. Semiconductor Industry Association, *2015 International Technology Roadmap* Executive Report, 6 (emphasis in original). See also *FCP*, 111–112 and Ceruzzi, *Computing*, 158. The ongoing densification of memory chips is not without its own problems, however (see note 132 in chapter 5).

54. Waldrop, "More than Moore," 146. On memory see also Katherine Bourzac, "Has Intel Invented a Universal Memory Tech?" *IEEE Spectrum*, April 19, 2017, https://spectrum.ieee.org/has-intel-invented-a-universal-memory-tech, accessed July 27, 2025. Foveros technology, after being introduced in the since-discontinued low-power 2018 10nm chip, only resurfaced in Intel's late-2023 product line (see Gareth Halfacree, "Intel's Foveros Tech Hits a Speedbump as Lakefield Gets Canned—One Year After Launch," *The Register*, July 8, 2021, https://www.theregister.com/2021/07/08/intel_lakefield_discontinued/, Paul Alcorn, "Intel Tapes In 7nm Meteor Lake Compute Tile," *Tom's Hardware*, May 24, 2021, https://www.tomshardware.com/news/intel-tapes-in-7nm-meteor-lake, and Gavin Bonshor, "Intel Releases Core Ultra H and U-Series Processors: Meteor Lake Brings AI and Arc to Ultra Thin Notebooks," *Anandtech*, December 14, 2023, https://www.anandtech.com/show/21185/intel-releases-core-ultra-h-and-u-series-processors-meteor-lake-brings-ai-and-arc-to-ultra-thin-notebooks; URLs accessed July 19, 2024).

55. This is a well-known state of affairs in engineering circles; see, e.g., *FCP*, 59 and box 2.3 (70–71).

56. This is an aspect of the "von Neumann bottleneck," after the Princeton architecture computing model spearheaded by John von Neumann, in which data and commands travel to the CPU along the same path (see John von Neumann, "First Draft of a Report on the EDVAC" (United States Army Ordnance Department Contract W-670-ORD-4926), Moore School of Electrical Engineering, University of Pennsylvania (June 30, 1945)). As Hillis notes, from the end of the Second World War, "computers were designed to keep the expensive processor switches as usefully busy as possible, while not placing too much demand on the speed of the memory" (Hillis, *The Pattern on the Stone*, 107). On the relative speeds of CPUs and memory, see 108–109 and *FCP*, e.g., 62–63, 105, 111–114. On the relative speeds of different types of memory and storage see Bourzac, "Has Intel Invented a Universal Memory Tech?" The NRC report also argues that "although" checks on Moore's law "pose major technical challenges, they will also drive considerable innovation in computing by forcing a rethinking of the [sequential] von Neumann model that has prevailed since the 1940s" (*FCP*, 11; see also 140–141 and Bourzac, "Has Intel Invented a Universal Memory Tech?"). Hillis argues that this architectural separation of processing and memory, perpetuated by the fact that "the speed of computers tended to double every two years," means "we are approaching the limits of how much we can speed up the computer without changing the basic design" (Hillis, *The Pattern on the Stone*, 108, 109).

57. See Intel Corporation, "Intel and Micron Produce Breakthrough Memory Technology" news release July 28, 2015, archived at https://web.archive.org/web/20200814145853/https://newsroom.intel.com/news-releases/intel-and-micron-produce-breakthrough-memory-technology/; Chaim Gartenberg, "Intel's New Optane SSDs are Superfast and Can Even Work as Extra RAM," *The Verge*, October 31, 2017, https://www.theverge.com/circuitbreaker/2017/10/31/16582018/intel-optane-p900-ssd-fast-dram-nand-flash-memory-desktop-computer (URLs accessed July 27, 2025); Bourzac, "Has Intel Invented a Universal Memory Tech?"; and Wendy Hui Kyong Chun, *Programmed Visions: Software and Memory* (Cambridge, MA: MIT Press, 2010), 137–173.

58. As the NRC report argues, "computing performance encompasses more than intrinsic CPU speed, but CPU performance has historically driven everything else: input/output, memory sizes and speeds, buses and interconnects, networks, and so on. If continued growth in CPU performance is threatened, so are the rest" (*FCP*, 11n2; see also 59). Collapsing memory and storage is one of their potential non-CPU-related recommendations (box 5.1 (135)).

59. A near-comprehensive collection of complaints about the 2016 model can be found at Michael Tsai, "New MacBook Pros and the State of the Mac," blog post

October 27, 2016, https://mjtsai.com/blog/2016/10/27/new-macbook-pros-and-the-state-of-the-mac/, accessed April 16, 2017.

60. Mike Wuerthele, "Intel's Chip Design, Not Apple's Choices, Reason Behind Thunderbolt 3 & RAM Issues in New MacBook Pro," *AppleInsider*, October 31, 2016, http://appleinsider.com/articles/16/10/31/intels-chip-design-not-apples-choices-reason-behind-thunderbolt-3-ram-issues-in-new-macbook-pro, accessed March 5, 2017; Dave James, "Intel Insider Claims It Finally Lost Apple Because Skylake QA 'Was Abnormally Bad,'" *PC Gamer*, June 24, 2020, https://www.pcgamer.com/intel-skylake-why-apple-left/, accessed June 25, 2020.

61. Apple Inc., "Apple Announces Mac Transition to Apple Silicon," June 22, 2020, https://www.apple.com/newsroom/2020/06/apple-announces-mac-transition-to-apple-silicon/, accessed June 22, 2020 (emphasis added). See also *FCP*, 69. For an analysis of Apple's move into custom silicon via its mobile products, see Pesce, "Zombie Moore's Law Shows Hardware Is Eating Software."

62. Florian Cramer, "What Is 'Post-digital'?" *APRJA* 3, no. 1 (2014), https://aprja.net/article/view/116068/165295, accessed April 25, 2020.

63. Andy Hertzfeld, *Revolution in the Valley: The Insanely Great Story of How the Mac Was Made* (Sebastopol, CA: O'Reilly Media, 2005).

64. Citing Waldrop and noting the correspondingly increasing expense of semiconductor fabrication (see Waldrop, "More than Moore," 147, but also *FCP*, 78 and Moore, "Moore's Law at 40," 81–82), Sean Cubitt argues that "venture capital, since the global financial crisis, has been hard put to find investors interested in risk. . . . The end product of this has been a diminution of invention, [and] a shift toward investing in applications, which have a better risk-to-profit ratio" (Sean Cubitt, *Finite Media: Environmental Implications of Digital Technologies* (Durham, NC: Duke University Press, 2017), 93). The essence of this shift is captured in the title of another of Chun's books: Wendy Hui Kyong Chun, *Updating to Remain the Same: Habitual New Media* (Cambridge, MA: MIT Press, 2016). See also note 92 later.

65. Anna Weiner, "It's Getting Harder to Believe in Silicon Valley," *The Atlantic*, March 2017, https://www.theatlantic.com/magazine/archive/2017/03/the-shine-comes-off-silicon-valley/513815/ accessed March 31, 2017. The book in question is Alexandra Wolfe, *Valley of the Gods: A Silicon Valley Story* (New York: Simon & Schuster, 2017). I consider the Valley's geographic and economic dynamics (as opposed to the cultural perception of its output) in chapter 5.

66. Laboria Cuboniks, *The Xenofeminist Manifesto: A Politics for Alienation* (2015), (London: Verso, 2018), 47 (§0x0C). See also Mark Alizart, *Cryptocommunism*, trans. Robin Mackay (Cambridge, UK: Polity, 2020), 25.

67. See, e.g., Goffey, "Algorithm," 15–17 and Hillis, *The Pattern on the Stone*, 78.

68. Kittler, "There Is No Software," 82.

69. Casey Alt, "Objects of Our Affection: How Object Orientation Made Computers a Medium," in *Media Archaeology: Approaches, Applications, and Implications*, eds. Erkki Huhtamo and Jussi Parikka (Berkeley: University of California Press, 2011), 279.

70. Ibid., 280.

71. Ibid., 290.

72. Ibid., 294; see also 291–292.

73. Ibid., 280. Theorists have noted more broadly the ways in which new media seems to emphasize space over time (e.g., Lev Manovich, *The Language of New Media* (Cambridge, MA: MIT Press, 2001), 157, and Chun, *Programmed Visions*, 3, 21). I have previously addressed the 1960s and their relationship to technological developments, postmodernism, and poststructural theory, alongside many others, in *The Global Sixties in Sound and Vision: Media, Counterculture, Revolt*, ed. Timothy Scott Brown and Andrew Lison (Basingstoke, UK: Palgrave Macmillan, 2014).

74. Lison, "New Media, 1989."

75. Alt, "Objects of Our Affection," 289, 296.

76. On late binding, see ibid., 294.

77. Ibid., 292.

78. Alan Kay quoted in ibid., 286 and 300n20. Kay's dissertation cites two of Engelbart's papers (Alan Curtis Kay, "The Reactive Engine," PhD diss. (University of Utah, 1969)). For a more recent consideration of Kay's work on Smalltalk and object-oriented programming, see Jacob Gaboury, *Image Objects: An Archaeology of Computer Graphics* (Cambridge, MA: MIT Press, 2021), 125–155.

79. Alan C. Kay, "The Early History of Smalltalk," *HOPL-II: The Second ACM SIGPLAN Conference on History of Programming Languages, ACM SIGPLAN Notices* 28, no. 3 (March 1993), 72 (emphases in original).

80. Kay's language here echoes both the postmodern fetishization of nonlinear complexity and Alt's claims for the effects of OOP: "The advent of a real Smalltalk on a real machine started off an explosion of parallel paths that are too difficult to intertwine in strict historical order" (Ibid., 79).

81. Ibid., 70.

82. For similar charts using different, older data see *FCP*, 9 (figure S.1), 91 (figure 3.3), and 156–158 (figures A.1–4) as well as the graphs in Sutter, "The Free Lunch Is Over" and Waldrop, "More than Moore," 146. The latest generations of CPUs can temporarily break past this long-standing barrier, but only in unsustainable bursts. Their standard clock rates remain limited.

83. On all of the above developments from 2004 onward, see *FCP*, e.g., 8–9, 11, 68, box 2.4 (72–73), 106–109, and appendix A (155–159); Waldrop, "More than Moore," e.g., 146; Sutter, "The Free Lunch Is Over"; and the summarization of Gelsinger's remarks in Brock, "Reflections on Moore's Law," 103.

84. Sutter, "The Free Lunch Is Over." Sutter uses "concurrency" and "parallelization" more or less interchangeably. That is, he employs it to describe all methods of parceling out clusters of instructions, from concurrency as described earlier to methods often gathered in contradistinction to it under the nomenclature of parallelization. These include virtual multiprocessing systems like hyperthreading, the multicore design of multiple processors on a single die, and multiple, fully distinct CPUs within the same system. For a similar, although not identical, taxonomy of parallelism, transitioning to which is the major theme of the report's recommendations, see *FCP*, 61. For an in-depth examination of parallelization see 105–131. For lay-accessible treatments, see Hillis, *The Pattern on the Stone*, 107–120 and the summary of computer scientist Rodney Brooks' remarks in Brock, "Reflections on Moore's Law," 105. Nevertheless, Sutter also makes clear that these various architectures comprise a hierarchy of performance and complexity and, moreover, that best-case parallelization scenarios must still sacrifice some efficiency: "Even having two threads running on two physical processors doesn't mean getting two times the performance" (Sutter, "The Free Lunch Is Over"). On this last point see also Waldrop, "More than Moore," 146 as well as *FCP*, 64–65 and Hillis, *The Pattern on the Stone*, 112–113, both describing the computer science principle known as Amdahl's law which establishes this (discussed further here in chapter 4).

85. Sutter, "The Free Lunch Is Over" (emphasis in original); Gelsinger summarized in Brock, "Reflections on Moore's Law," 104. Increased parallel-processing efficiency in the OS and other levels of the software development "stack" is another key recommendation of *FCP*, e.g., 14–15, 127–130, 137–138, 141.

86. See the Squeak/Smalltalk wiki, "Thinking About Massively Parallel Smalltalk," July 4, 2007, http://wiki.squeak.org/squeak/537, accessed April 16, 2017.

87. Alt, "Objects of Our Affection," 299.

88. *FCP*, Box 2.3 (71). See also Kittler, "There Is No Software," 89.

89. E.g., the assertion that "humanities departments need intensive computing power to keep up their research in today's era of large-scale text and data sets, distributed archival resources, and multi-modal (visual, aural, cartographic, etc.) materials" in Alan Liu and William G. Thomas III, "Humanities in the Digital Age" (2012/2015) in *Between Humanities and the Digital*, ed. Patrik Svensson and David Theo Goldberg (Cambridge, MA: MIT Press, 2015), 38. Liu and Thomas pose the adoption of the digital humanities within the university as a response to austerity policies offering departments and faculty some control over these ostensibly inevitable trends. For critiques of the digital humanities' uncritical embrace of technology

and its complicity with neoliberalism, see, e.g., Kathryn Conrad, "What the Digital Humanities Can't Do," *The Chronicle of Higher Education*, September 8, 2014, https://www.chronicle.com/article/what-the-digital-humanities-cant-do/, accessed July 27, 2025 and Daniel Allington, Sarah Brouillette, and David Golumbia, "Neoliberal Tools (and Archives): A Political History of Digital Humanities," *Los Angeles Review of Books*, May 1, 2016, https://lareviewofbooks.org/article/neoliberal-tools-archives-political-history-digital-humanities/, accessed April 15, 2017, respectively.

90. Although language developers are adding such capabilities, they have not been historically significant from a DH perspective. For example, Python 3, which only fully supplanted prior versions in 2020, made its _thread API nonoptional only with version 3.7 (see https://docs.python.org/3/library/_thread.html); R, meanwhile, has a similar "parallel" package, but at a low level the language "was designed to use only a single thread (processor) at a time. Even today, R works that way unless linked with multi-threaded BLAS/LAPACK libraries" (Microsoft R Application Network, "The Benefits of Multithreaded Performance with Microsoft R Open," archived at https://web.archive.org/web/20220508022024/https://mran.microsoft.com/documents/rro/multithread) (URLs accessed July 27, 2025).

91. In the series of lectures mentioned earlier, Kittler states that, with the end of Moore's law, "nature will not look into her own eyes anymore but into muddy waters which don't reflect anything" (Kittler, "Computing, Ubiquity and the One Single Medium"). For readings of Moore's law, its breakdown, and their relationship to environmental crises, see Anne Pasek, "Seeing Carbon through Silicon," *Logic* 9 (December 7, 2019), https://logicmag.io/nature/seeing-carbon-through-silicon/ accessed June 11, 2020 and "From Atoms to Electrons: An Energy History and Future of Computing" in *Digital Energetics*, ed. Anne Pasek et al. (Lüneburg, Germany/Minneapolis: meson/University of Minnesota Press, 2023), 15–42. In the former piece, alongside the physical, Pasek, like Waldrop, highlights human factors enabling Moore's law, e.g., "considerable barn-raising by key figures in the industry" (Pasek, "Seeing Carbon Through Silicon") ultimately codified in the SIA. By contrast, she argues, the populist aspect of climate justice is excluded in technocratic efforts to establish a similar decarbonization "law." As she herself notes, however, "the risk here, as with any framing comparison, is that the metaphor will not hold" (ibid.), and the SIA, for example, is hardly an activist organization. The failures (or successes) of attempts to technologically bring about an exponential reduction in emissions will be largely independent of those to perpetuate Moore's law. See also *FCP*, 142–145 on the environmental impact of computing's energy usage, including how it can diminish utilization elsewhere in the economy. I discuss the interrelationship of computation, energy, and waste further in chapter 3.

92. Alex Williams and Nick Srnicek, "#Accelerate: Manifesto for an Accelerationist Politics," (2013) in *#ACCELERATE#: The Accelerationist Reader*, ed. Robin Mackay and Armen Avanessian (London: Urbanomic, 2014), 355, 362. See also Marx, "Preface

to *A Contribution to the Critique of Political Economy*." The more outlandish claims of accelerationism vis-à-vis technological progress are explored at length in chapter 5.

93. Cuboniks, *Xenofeminist Manifesto*, 75 (§0x13).

94. Conrad, "The Price of Programmability," 266.

95. Wendy Hui Kyong Chun, "The Enduring Ephemeral, or the Future Is a Memory," *Critical Inquiry* 35, no. 1 (Autumn 2008): 170. The limit represented by absolute processing power utilization is mirrored economically in not only the Marxist concept of full employment but also that of monolithic industry. As Mandel writes, for Marx, "totally monopolized capital . . . would essentially be stagnating capital" (Mandel, "Introduction," 60).

96. "All code operations, despite such metaphoric faculties as call or return, come down to absolutely local string manipulations, that is, I am afraid, to signifiers of voltage differences" (Kittler, "There Is No Software," 84).

97. After the characterization of the ultimate determinism of economic factors in Louis Althusser, "Contradiction and Overdetermination" (1962) in *For Marx* (1965) trans. Ben Brewster (1969) (London: Verso, 2005), 111. Although the conclusions he draws from this are much more expansive, Althusser borrows this notion from a late letter of Engels, reprinted in English as "Engels to J. Bloch in Königsberg, September 21 [–22], 1890" in Karl Marx, Frederick Engels, and Vladimir Lenin, *On Historical Materialism*, compiled by T. Borodulina (Moscow: Progress Publishers, 1972), 294, where it is rendered as "the *ultimately* determining element in history is the production and reproduction of real life" (italics in original).

98. Ceruzzi, *Computing*, 159.

99. These developments are discussed further in chapter 3. Sutter suggests that in the wake of static clock rates "*efficiency and performance optimization will get more, not less, important*. Those languages that already lend themselves to heavy optimization will find new life; those that don't will need to find ways to compete and become more efficient and optimizable. Expect long-term increased demand for performance-oriented languages and systems" (Sutter, "The Free Lunch Is Over" (emphasis in original)); Waldrop, "More than Moore," 145; Pesce, "Zombie Moore's Law Shows Hardware is Eating Software." See also the summary of Brooks's remarks in Brock, "Reflections on Moore's Law," 105; and *FCP*, e.g., 100–103, 115, 127–130, 138–141, 151. *FCP* argues that "the progress of Moore's law hid the costs of abstraction" layers (107; see also 71 and 106–109).

Chapter 2

1. For a view of the industry from this perspective see Sean Cubitt, *Finite Media: Environmental Implications of Digital Technologies* (Durham, NC: Duke University Press,

2017), 86–95. This does not take into account the extractive, energy, e-waste, and other environmental impacts associated with digital mediation covered throughout the rest of Cubitt's book, either.

2. That said, the economics of semiconductor assembly do make an appearance in chapter 5 as part of my consideration of Silicon Valley labor dynamics.

3. Fredric Jameson, "Three Names of the Dialectic" in *Valences of the Dialectic* (London: Verso, 2009), 46. See also Fredric Jameson, *Representing* Capital*: A Reading of Volume One* (London: Verso, 2011), 55–56 and notes 6 and 18 in chapter 1.

4. Gordon E. Moore, "Moore's Law at 40," in *Understanding Moore's Law: Four Decades of Innovation*, ed. David C. Brock (Philadelphia: Chemical Heritage Press, 2006), 67.

5. David C. Brock, "Reflections on Moore's Law" in *Understanding Moore's Law*, ed. Brock, 97.

6. See, e.g., M. Mitchell Waldrop, "More than Moore," *Nature* 530 (February 11, 2016): 147.

7. Gordon E. Moore, "Cramming More Components onto Integrated Circuits" (1965), reprinted in *Proceedings of the IEEE* 86, no. 1 (January 1998): 83. "Curve" here indicates the increased cost per component of including either fewer or more of them relative to this minimum.

8. For a more through consideration of yield, see Gordon E. Moore, "Semiconductor Integrated Circuits," in *Microelectronics: Theory, Design, and Fabrication*, ed. Edward Keonjian (New York: McGraw-Hill, 1963), 347–353 (discussed further later).

9. See, e.g., Intel, "Intel® Processor Names and Numbers," https://www.intel.com/content/www/us/en/processors/processor-numbers.html, accessed July 28, 2025.

10. See, e.g., Scharon Harding, "What Is Binning? A Basic Definition," *Tom's Hardware*, September 12, 2022, https://www.tomshardware.com/reviews/glossary-binning-definition,5892.html and Andrew Cunningham, "Report: Apple Buys Every 3 nm Chip That TSMC Can Make for Next-gen iPhones and Macs," *Ars Technica*, August 7, 2023, https://arstechnica.com/gadgets/2023/08/report-apple-is-saving-billions-on-chips-thanks-to-unique-deal-with-tsmc/ (both URLs accessed July 31, 2024).

11. Cerebras, "Wafer-Scale Processors: The Time Has Come," Cerebras blog, September 26, 2019, archived at https://web.archive.org/web/20240620225727/https://www.cerebras.net/blog/wafer-scale-processors-the-time-has-come/, accessed July 28, 2025.

12. Correspondingly, as suggested in the introduction, the theoretical condition of 100% utilization could thus be further out than the end of Moore's law itself, though they are intimately related.

13. This realization goes back to the early days of microelectronics; see, e.g., Moore, "Integrated Semiconductor Circuits," 349.

14. Cerebras, "Wafer-Scale Processors."

15. Cerebras, "Wafer-Scale Engine 3: The Largest Chip Ever Built," press release, 2024, https://8968533.fs1.hubspotusercontent-na1.net/hubfs/8968533/Datasheets/WSE-3%20Datasheet.pdf, accessed July 29, 2024.

16. Sebastian Moss, "Cerebras Unveils Four Trillion-Transistor Giant Chip, Targets Generative AI," *Data Center Dynamics*, March 13, 2024, https://www.datacenterdynamics.com/en/news/cerebras-unveils-four-trillion-transistor-giant-chip-targets-generative-ai, accessed July 29, 2024.

17. Cerebras, "Cerebras CS-3: The World's Fastest and Most Scalable AI Processor," Cerebras blog, March 12, 2024, https://www.cerebras.net/blog/cerebras-cs3, accessed July 29, 2024.

18. Ibid.

19. Toby Sterling, "TSMC Says Can Make Next Generation Chips Without ASML's New Machine," *Reuters*, May 14, 2024, https://www.reuters.com/technology/tsmc-says-asml-high-na-euv-machine-not-necessary-a16-node-2024-05-14/, accessed August 1, 2024.

20. See, e.g., the chart from International Business Strategies, a technology consulting firm, in Cerebras, "Wafer-Scale Processors."

21. See, e.g., Yawen Ran, Yu Yang, Yun Wang, and Yi Liu, "Dynamics of the Global Semiconductor Trade and Its Dependencies," *Journal of Geographical Sciences* 33, no. 6 (2023): 1141–1160.

22. Andre Barbe, Dan Kim, and David Riker, "Trade and Labor in the U.S. Semiconductor Industry," *Journal of International Commerce and Economics* (July 2018): 4; available online at https://www.usitc.gov/publications/332/journals/barbe_kim_and_riker_-_trade_and_labor_in_the_us_semiconductor_industry_.pdf, accessed July 28, 2024. See also Ran et al., "Dynamics of the Global Semiconductor Trade," e.g., 1148.

23. Barbe et al., "Trade and Labor in the U.S. Semiconductor Industry," 3, 5–10.

24. Cerebras, "Cerebras Systems Unveils World's Fastest AI Chip with Whopping 4 Trillion Transistors," press release, March 14, 2024, https://www.cerebras.net/press-release/cerebras-announces-third-generation-wafer-scale-engine; Taiwan Semiconductor Manufacturing Company, "Logic Technology," https://www.tsmc.com/english/dedicatedFoundry/technology/logic (both URLs accessed July 31, 2024).

25. Cunningham, "Report"; Ben Lovejoy, "Secret Meeting Between Apple and TSMC Reported, Possibly to Reserve All 2nm Capacity," *9to5Mac*, May 20, 2024, https://9to5mac.com/2024/05/20/secret-meeting-between-apple-and-tsmc-reported-possibly-to-reserve-all-2nm-capacity/; Tim Hardwick, "Apple Secures Half of TSMC's

2nm Production Capacity for iPhone 18," *MacRumors*, August 28, 2025, https://www.macrumors.com/2025/08/28/apple-tsmc-2nm-production-iphone-18/ (both URLs accessed September 11, 2025.)

26. Cunningham, "Report."

27. Anton Shilov, "Intel Foundry Unit Loses $7 Billion in 2023, Company Outsources 30% of its Production to TSMC, Others," *Tom's Hardware*, April 3, 2024, https://www.tomshardware.com/tech-industry/intel-foundry-unit-loses-dollar7-billion-in-2023-company-outsources-30-of-its-production-to-tsmc-others, accessed July 28, 2024.

28. See, e.g., Don Clark, Tripp Mickle, and Steve Lohr, "Intel's Chief Executive Is Out Amid Chipmaker's Struggles," *New York Times*, December 2, 2024, https://www.nytimes.com/2024/12/02/technology/intel-ceo-pat-gelsinger.html, accessed January 13, 2025.

29. Lovejoy, "Secret Meeting." As the article on Gelsinger's ouster notes, its yields are significantly higher: "The difference in the production quality between TSMC and Intel means that customers would have to spend more money with Intel to produce the same number of chips" (Clark et al., "Intel's Chief Executive Is Out").

30. Taiwan Semiconductor Manufacturing Company Limited, "Parent Company Only Financial Statements for the Years Ended December 31, 2023 and 2022 and Independent Auditors' Report," available online at https://investor.tsmc.com/sites/ir/financial-report/2023/TSMC%202023Q4%20Unconsolidated%20Financial%20Statements_E.pdf, accessed July 28, 2024, 115, 30. Figures are unconsolidated, and thus do not include subsidiaries.

31. Ibid., 117, 116.

32. NVIDIA, 2024 Form 10-K, available online at https://investor.nvidia.com/financial-info/sec-filings/sec-filings-details/default.aspx?FilingId=17293267, accessed July 28, 2024, 40, 69. Some caveats: Unlike TSMC's, NVIDIA's figures are consolidated. NVIDIA's year ends January 28, 2024, while TSMC's closes on December 31, 2023. R&D operating expenses undoubtedly include more than wages, meaning this ratio is an upper limit. Additionally, TSMC uses International Financial Reporting Standards while NVIDIA applies Generally Accepted Accounting Principles, further complicating an apples-to-apples comparison. Again, however, the difference is so stark as to render these divergences mostly irrelevant.

33. Cubitt, *Finite Media*, 83–84. For more on these dynamics, see Lisa Nakamura, "Indigenous Circuits: Navajo Women and the Racialization of Early Electronic Manufacture," *American Quarterly* 66, no. 4 (December 2014): 919–941. I discuss this piece, as well as Marx's analysis of the tendency of the profit rate to fall, in chapter 5.

34. Barbe et al., "Trade and Labor in the U.S. Semiconductor Industry," 3. The authors' main objective is to chart a domestic "influx of workers into higher-paying

occupations (mostly engineering jobs)" (11) alongside both an absolute decline in the number of US industry positions and a consequent outflow of lower-wage manufacturing jobs from the country to the rest of the world. Interestingly, they are able make this case even without taking design-only companies into account. Doing so, they write, "would further highlight this impact" (13n24).

35. Ibid., 3. See also Cubitt, *Finite Media*, 87, 93–94. Cubitt's analysis often conflates fabrication and assembly, as does Nakamura at times.

36. Google DeepMind, "How AlphaChip Transformed Computer Chip Design," September 26, 2024 blog post, https://deepmind.google/discover/blog/how-alphachip-transformed-computer-chip-design/, accessed January 13, 2025.

37. Karl Marx, *Capital: A Critique of Political Economy*, vol. 1 (1867) trans. Ben Fowkes (1976) (London: Penguin, 1990), 509.

38. Ibid., 510.

39. Arnold Thackray, David C. Brock, and Rachel Jones, *Moore's Law: The Life of Gordon Moore, Silicon Valley's Quiet Revolutionary* (New York: Basic Books, 2015), 230.

40. Moore, "Semiconductor Integrated Circuits," 266. See also Thackray et al., *Moore's Law*, 230.

41. Ibid., 232. See also note 91 in chapter 1.

42. From the Latin for "workplace" (*Oxford English Dictionary*, s.v. "laboratory (*n.*)").

43. Moore, "Semiconductor Integrated Circuits," 266.

44. Ibid., 343. For a condensed list and explication see 338–342.

45. Ibid., 344–345.

46. Ibid., 343–344.

47. David C. Brock notes that the mid-1950s concept for Shockley Semiconductor relied on the way "diffusion lent itself to mass production, since it was inherently a chemical batch process in which many device structures were created on multiple wafers at the same time. If diffusion batch processing could be automated, there was the possibility for significantly lowering the production cost of transistors. . . . Diffusion allowed engineers to create thinner layers of chemically distinct layers in silicon, and these thinner layers meant faster transistors" (David C. Brock, "From Automation to Silicon Valley: The Automation Movement of the 1950s, Arnold Beckman, and William Shockley," *History and Technology* 28, no. 4 (December 2012): 391).

48. Moore, "Semiconductor Integrated Circuits," 301, 304.

49. Ibid., 325.

50. Ibid., 326.

51. Ibid., 350

52. Marx, *Capital* vol. 1, 510–511.

53. Ibid., 515.

54. Sterling, "TSMC Says Can Make Next Generation Chips Without ASML's New Machine."

55. Barbara Ortutay, "Chipmaker Intel to Cut 15,000 Jobs as Tries to Revive Its Business and Compete with Rivals," *Associated Press*, August 1, 2024, https://apnews.com/article/intel-chip-ai-job-cuts-layoffs-loss-e61781e9364b69af63481c34ca5dcd67, accessed August 4, 2024; Ashley Lutz, "Intel Plans to Slash 25,000 Jobs in 2025 as New CEO Warns, 'There Are No More Blank Checks,'" *Fortune*, July 25, 2025, https://fortune.com/2025/07/25/intel-layoffs-ceo-staff-memo-stock-price/, accessed July 29, 2025. Ironically, for the latter story, "*Fortune* used generative AI to help with an initial draft."

56. Jameson, *Representing* Capital, 56.

57. These obstacles go beyond the (as I have been arguing, temporary) barriers presented by cost. As the National Research Council report notes, "If parallel programming and related software efforts fail to become widespread, the development of exciting new applications that drive the computer industry will stall; if such innovation stalls, many other parts of the economy will follow suit" (The National Research Council Committee on Sustaining Growth in Computing Performance, *The Future of Computing Performance: Game Over or Next Level?*, ed. Samuel H. Fuller and Lynette I. Millett (Washington, DC: National Academies Press, 2011), 5), to which it is only necessary to add, obviously and dialectically, that if the economy stalls, so, too, will technological innovation.

58. David Harvey, *A Brief History of Neoliberalism* (2005) (Oxford: Oxford University Press, 2007), 159.

59. Sometimes also called the "Solow paradox" after its succinct formulation in a book review of Stephen S. Cohen and John Zysman, *Manufacturing Matters: The Myth of the Post-Industrial Economy* (New York: Basic Books, 1987): "You can see the computer age everywhere but in the productivity statistics" (Robert M. Solow, "We'd Better Watch Out," *New York Times*, July 12, 1987, 36). See also Aaron Benanav, *Automation and the Future of Work* (London: Verso, 2019), 17–18 (discussed further in chapter 5).

60. Erik Brynjolfsson, "The Productivity Paradox of Information Technology," *Communications of the ACM* 36, no. 12 (December 1993): 75.

61. Harvey, *A Brief History of Neoliberalism*, 19.

62. Nick Srnicek, *Platform Capitalism* (Cambridge, UK: Polity, 2017), 9–33.

63. Ibid., 39.

64. Ibid., 45, 64–70, 116.

65. Ibid., 45.

66. Ibid., 86.

67. Ibid., 40.

68. Ibid., 97.

69. See Marx, *Capital* vol. 1, 533–542.

70. Jonathan Beller, *The World Computer: Derivative Conditions of Racial Capitalism* (Durham, NC: Duke University Press, 2021), 113. Similarly, Adam Greenfield writes, contemporary monopolistic tech platforms pursue "the same goal: to mediate and monetize everyday life to the maximum possible extent" (Adam Greenfield, *Radical Technologies: The Design of Everyday Life* (London: Verso, 2017), 283).

71. For an in-depth treatment of this expansion see Jonathan Crary, *24/7: Late Capitalism and the Ends of Sleep* (London: Verso, 2013). For Crary, it cuts across speculative markets and more basic means of value extraction: "Whether one's vantage point is Marxist or not, there is no evading the extent to which the internet and digital communications have been the engine of the relentless financialization *and commodification* of more and more regions of individual and social life" (99, emphasis mine). However, he also emphasizes the perpetual self-revolutionizing of capitalist digital technology that an end to Moore's law places into question (e.g., 37). See also Greenfield, *Radical Technologies*, 32.

72. Edward Nik-Khah and Philip Mirowski, "The Ghosts of Hayek in Orthodox Microeconomics: Markets as Information Processors" in Armin Beverungen, Edward Nik-Khah, Philip Mirowski, and Jens Schröter, *Markets* (Lüneburg, Germany/Minneapolis: meson/University of Minnesota Press), 61.

73. Nik-Khah and Mirowski are responding to a prominent market design economist's assertion that "if you really want to operate at digital speeds, you need to take people out of the middle of the process" (Alvin E. Roth, *Who Gets What and Why: The New Economics of Matchmaking and Market Design* (Boston: Houghton Mifflin Harcourt, 2015), 101). On HFT specifically, see Alberto Toscano, "Gaming the Plumbing: High-Frequency Trading and the Spaces of Capital," *Mute* 3, no. 4 (January 2013), available online at https://www.metamute.org/editorial/articles/gaming-plumbing-high-frequency-trading-and-spaces-capital, accessed March 22, 2023.

74. Beller, *The World Computer*, 116, 117.

75. Marc Steinberg, *The Platform Economy: How Japan Transformed the Consumer Internet* (Minneapolis: University of Minnesota Press, 2019), 6. Steinberg traces the rise of platform discourse in Japan in part to a sense that hardware was destined to become

ever-cheaper (15). This is, of course, a consequence of Moore's law. Cubitt similarly argues that "financialization is datafication" (Cubitt, *Finite Media*, 79).

76. Stephen Roach, "Services Under Siege—The Restructuring Imperative," *Harvard Business Review* September–October 1991, 81–92. See also Brynjolfsson, "The Productivity Paradox of Information Technology," 75.

77. Daron Acemoglu, David Autor, David Dorn, Gordon H. Hanson, and Brendan Price, "Return of the Solow Paradox? IT, Productivity, and Employment in US Manufacturing," *American Economic Review* 104, no. 5 (May 2014): 394–399. For a more skeptical view that cites this study, see note 103 in Chapter 5.

78. Srnicek would seem to agree: "Whereas the tech boom of the 1990s at least left us with the basis for the internet, the tech boom of the 2010s looks as though it will simply leave us with premium services for the rich" (Srnicek, *Platform Capitalism*, 121). Given its history goes back to the late 1960s, much hinges on what is meant by "the basis for the internet," but the commercialized Internet that twenty-first-century platforms leverage did begin to take form in the 1990s.

79. See Lison, "New Media, 1989."

80. As noted above, see, e.g., Steinberg, *The Platform Economy*, 46 and 74. His examples come from Japan, but are applicable to nearly all personal computing of the era.

81. Harvey, *A Brief History of Neoliberalism*, 159.

82. Jonathan Sterne, *MP3: The Meaning of a Format* (Durham, NC: Duke University Press, 2012), e.g. 208–209, 215, 217, and figures 30–31 on 218–219.

83. Ibid., 194. See also Philip Sherburne, "Digital DJing App That Pulls You In," *Grooves* 10 (2003), 46–47. The classic work on this development more broadly is N. Katherine Hayles, *How We Became Posthuman: Virtual Bodies in Cybernetics, Literature, and Informatics* (Chicago: University of Chicago Press, 1999).

84. By definition, the larger a stream of information is, the more storage required to hold it. Conversely, the more efficiently it is compressed, the more processing power necessary to expand it. This relationship is dialectical: "Rather perversely it happens that as one makes the coding more and more nearly ideal, one is forced to longer and longer delays *in the process of coding*" (Warren Weaver, "Recent Contributions to the Mathematical Theory of Communication," in Claude E. Shannon and Warren Weaver, *The Mathematical Theory of Communication* (1948/1949) (Urbana: University of Illinois Press, 1998), 18 (emphasis in original)). See also notes 92 and 93 in chapter 3.

85. For an enlightening discussion of this process vis-à-vis the smartphone, see Greenfield, *Radical Technologies*, 9–29.

86. Brynjolfsson, "The Productivity Paradox of Information Technology," 73, 76.

87. Erik Brynjolfsson and Avinash Collis, "How Should We Measure the Digital Economy?" *Harvard Business Review* November–December 2019, available online at https://hbr.org/2019/11/how-should-we-measure-the-digital-economy, accessed March 21, 2023. Wikipedia vs. the no-longer-existent *Encyclopedia Britannica* set is one of their key examples.

88. Erik Brynjolfsson and Andrew McAfee, *The Second Machine Age: Work, Progress, and Prosperity in a Time of Brilliant Progress* (New York: W.W. Norton & Co., 2014). Their work is one of the key catalysts of the left-wing responses to automation discourse explored here at length in chapter 5.

89. Anne Pasek, "From Atoms to Electrons: An Energy History and Future of Computing" in *Digital Energetics*, ed. Anne Pasek et al. (Lüneburg, Germany/Minneapolis: meson/University of Minnesota Press, 2023), 17.

90. "We may compare a man in the process of computing a real number to a machine which is only capable of a finite number of conditions" (A. M. Turing, "On Computable Numbers, with an Application to the Entscheidungsproblem" (1936), *Proceedings of the London Mathematical Society* ser. 2, vol. 42 (1937): 231).

91. Ibid., 232, 241.

92. "His machine is intended to model what a human mathematician does when calculating or reasoning, and by showing what limits this machine encounters, he identifies constraints which bind mathematical reasoning in general (whether done by humans or machines)" (Paul Cockshott, "Turing: The Irruption of Materialism into Thought," *Oxford University Press* blog post June 19, 2012, https://blog.oup.com/2012/06/turing-the-irruption-of-materialism-into-thought/, accessed May 10, 2017). See also note 31 in chapter 1.

93. Norbert Wiener, *The Human Use of Human Beings: Cybernetics and Society* (1950) Revised Edition (1954) (Cambridge, MA: Da Capo Press, 1988), 150. See also Norbert Wiener, "Letter Covering the Memorandum on the Scope, etc. of a Suggested Computing Machine (September 21, 1940)" and "Memorandum on the Mechanical Solution of Partial Differential Equations" (1940) in Norbert Wiener, *Collected Works with Commentaries* volume IV, ed. Pesi Masani (Cambridge, MA: MIT Press, 1983), 122–134.

94. The *OED* lists the earliest use of "program" in relation to computers as John Mauchly's 1942 memo on the use of vacuum tubes, which places it after Wiener's stated revelation but prior to its publication in *The Human Use of Human Beings* (*OED*, s.v. "programme | program, *n.*"; see also note 48 in chapter 3). In computing, vacuum tubes were the predecessors to transistors.

95. Wiener, "Memorandum on the Mechanical Solution of Partial Differential Equations," 133.

96. Alan Turing, "Intelligent Machinery," National Physical Laboratory report (1948), 8. This is a connection other pioneers of the information age also made (see, e.g., Wiener, *The Human Use of Human Beings*, 124).

97. Konrad Zuse, "Calculating Space" (1969) trans. Aztec School of Languages (1970), in *A Computable Universe: Understanding and Exploring Nature as Computation*, ed. Hector Zenil (Hackensack: World Scientific, 2013), 730.

98. Srnicek, *Platform Capitalism*, 5. See also note 12 in chapter 1 and note 57 earlier.

Chapter 3

1. Mark Pesce, "Zombie Moore's Law Shows Hardware Is Eating Software," *The Register,* September 22, 2016, https://www.theregister.com/2016/09/22/the_evolution_of_moores_law_suggests_hardware_is_eating_software/, accessed July 27, 2025. See also note 99 in chapter 1.

2. A. M. Turing, "On Computable Numbers, with an Application to the Entscheidungsproblem" (1936), *Proceedings of the London Mathematical Society* ser. 2, vol. 42 (1937): 233–235; Andrew Goffey, "Intelligence" in *Software Studies: A Lexicon*, ed. Matthew Fuller (Cambridge, MA: MIT Press, 2008), 132.

3. Andrew Goffey, "Algorithm" in *Software Studies*, ed. Fuller (Cambridge, MA: MIT Press, 2008), 17. It is this first concept that Kittler builds upon when ruminating about future hardware (see note 36 in chapter 1).

4. See Fredric Jameson, "Three Names of the Dialectic," in *Valences of the Dialectic* (London: Verso, 2009), 3–70.

5. Friedrich Kittler, "There Is No Software," *Stanford Literature Review* 9.1 (Spring 1992): 84. For this type of reception see, e.g., Geoffrey Winthrop-Young, *Kittler and the Media* (Cambridge, UK: Polity, 2011), 75: "So bent is Kittler on exposing the hollowness of software that he came to deny its very existence." Other knowledgeable analyses also trade on this kind of sensationalism, e.g., Paul Feigelfeld, "Kittler Is a Liar!" interview (2013) trans. Daniel Bowles, *Metaphora: Journal for Literary Theory and Media*, Edited Vol. 1, eds. Arndt Niebisch and Martina Süess (2015), archived at https://web.archive.org/web/20160615101257/https://metaphora.univie.ac.at/volume1-feigelfeld-eng.pdf, accessed August 16, 2022.

6. Erkki Huhtamo and Jussi Parikka, "Introduction: An Archaeology of Media Archaeology," in *Media Archaeology: Approaches, Applications, and Implications*, ed. Erkki Huhtamo and Jussi Parikka (Berkeley: University of California Press, 2011), 8; Lev Manovich, *Software Takes Command* (New York: Bloomsbury, 2013).

7. Cf. Siegfried Giedion, *Mechanization Takes Command: A Contribution to Anonymous History* (New York: Oxford University Press, 1948). See also Manovich, *Software Takes Command*, 5.

8. Ibid., 1–2.

9. Ibid., 20, 21.

10. Ibid., 332.

11. Ibid., 337, 338.

12. Ibid., 4, 332.

13. Alan C. Kay, "The Early History of Smalltalk," *HOPL-II: The Second ACM SIGPLAN Conference on History of Programming Languages, ACM SIGPLAN Notices* 28, no. 3 (March 1993), 70. See also note 79 in chapter 1. For an analysis of Smalltalk, the GUI, and PARC from the perspective of software primacy, see Manovich, *Software Takes Command*, e.g., 56–58 and 96–104.

14. Alan C. Kay, "A Personal Computer for Children of All Ages," *ACM '72: Proceedings of the ACM Annual Conference*, Vol. 1 (August 1972), 8.

15. Alan Curtis Kay, "The Reactive Engine," PhD diss. (University of Utah, 1969), 226. It should be noted that the appeal to multiple chips in this and the previous quote does not contradict Smalltalk's single-processor focus outlined in chapter 1. What is being described here is the employment of multiple specialized and/or memory chips rather than general-purpose CPUs (see note 47 later).

16. For a detailed examination of both the logical and material relationships between computation and visual display, including the work of Kay and others at PARC, see Jacob Gaboury, *Image Objects: An Archaeology of Computer Graphics* (Cambridge, MA: MIT Press, 2021).

17. Alan Kay quoted in Owen W. Linzmayer, *Apple Confidential 2.0: The Definitive History of the World's Most Colorful Company* (San Francisco: No Starch Press, 2004), 78. As Linzmayer himself relates, "the Lisa was so slow that it inspired a popular 'knock knock' joke, wherein the response to 'Who's there?' was a 15-second pause followed by the word 'Lisa'" (ibid.). For a history of the relationship between Xerox and Apple see 73–83.

18. Ibid., 78. Kay claims of the 1978 prototype NoteTaker laptop, a stripped-down system modeled, like the Star, on the Xerox Alto, that "if Xerox (and PARC) as a whole had believed in these smaller scale ideas, we could have put much more silicon muscle behind the dreams and successfully built them in the 70's when they were first possible" (Kay, "The Early History of Smalltalk," 87), but this would have undoubtedly increased the cost in comparison with the commodity parts used in Kay's prototypes.

19. On smartphones see, e.g., Adam Greenfield, *Radical Technologies: The Design of Everyday Life* (London: Verso, 2017), 25. This kind of interface trickery is ongoing (see, e.g., Richard Nieva, "Google Buried Location Settings So People Wouldn't

Disable Them, Court Docs Claim," *CNET*, May 29, 2021, https://www.cnet.com/tech/mobile/google-buried-location-settings-so-people-wouldnt-disable-them-court-docs-claim/, accessed July 29, 2025). Ironically, key critiques of the GUI as ideological (e.g., Wendy Hui Kyong Chun, *Programmed Visions: Software and Memory* (Cambridge, MA: MIT Press, 2010), 55–95) are often more even-handed, identifying the potential they hold for both delusion *and* demystification.

20. See Andrew Lison, "New Media, 1989: Cubase and the New Temporal Order," *Computational Culture* 8 (July 2021), http://computationalculture.net/new-media-1989-cubase-and-the-new-temporal-order/.

21. This is implicit in Turing, "On Computable Numbers," but see also, e.g., W. Daniel Hillis, *The Pattern on the Stone: The Simple Ideas That Make Computers Work* (1998), 2nd ed. (New York: Basic Books, 2015), 62.

22. Pesce, "Zombie Moore's Law Shows Hardware Is Eating Software" (emphasis in original). Similarly, the National Research Council report argues that "in the future, the term *software stack* will be a misnomer. A 'software-hardware stack' will be the norm. The hardware, the programming model, and the applications will all need to change" (The National Research Council Committee on Sustaining Growth in Computing Performance, *The Future of Computing Performance: Game Over or Next Level?*, ed. Samuel H. Fuller and Lynette I. Millett (Washington, DC: National Academies Press, 2011), 128; hereafter referred to as *FCP*; emphasis in original).

23. For an example of this practice see the close reading of the Amiga game *Menace* (Psygnosis, 1988) in Jimmy Maher, *The Future Was Here: The Commodore Amiga* (Cambridge, MA: MIT Press, 2012), 224–245.

24. M. Mitchell Waldrop, "More than Moore," *Nature* 530 (11 February 2016): 145. See also *FCP*, 139.

25. RetroRGB, "MiSTer FPGA Hardware," https://www.retrorgb.com/mister.html, accessed June 19, 2021.

26. Arduino, "FPGA HDL Basics," https://docs.arduino.cc/learn/programming/vidor/, accessed July 30, 2025.

27. Ibid.

28. While bullish on custom silicon, the National Research Council argues that "continued innovation is needed to develop programming systems that can harness the potential parallelism of FPGAs" (*FCP*, 16).

29. Fabian Offert, "Hands on Circuits: Preserving the Semantic Surplus of Circuit-level Functionality with Programmable Logic Devices," in *Hands on Media History: A New Methodology in the Humanities and Social Sciences*, ed. Nick Hall and John Ellis (London: Routledge, 2020), 224.

30. Ibid., 227 (emphasis added).

31. Ibid., 225–226. Offert's citations here include Kittler, "There Is No Software."

32. See, e.g., note 42 in the introduction.

33. Alexander R. Galloway, *Protocol: How Control Exists After Decentralization* (Cambridge, MA: MIT Press, 2004), 165–166. See note 41 in the introduction for further discussion and Chun's critique.

34. Byuu, "FPGAs Aren't Magic," February 6, 2018 (updated December 27, 2019), archived at https://archive.ph/fWosI, accessed July 22, 2022.

35. Near makes similar points using the earlier Analogue Super Nt FPGA system (Byuu, "FPGAs Aren't Magic"). For an example of these cores, see https://github.com/MiSTer-devel/Distribution_MiSTer, accessed July 22, 2022.

36. Byuu, "Accuracy Takes Power: One Man's 3GHz Quest to Build a Perfect SNES Emulator," *Ars Technica* August 9, 2011 https://arstechnica.com/gaming/2011/08/accuracy-takes-power-one-mans-3ghz-quest-to-build-a-perfect-snes-emulator/, accessed July 22, 2022. See also Byuu, "FPGAs Aren't Magic."

37. RetroRGB, "MiSTer FPGA Hardware"; Offert, "Hands On Circuits," 229; Byuu, "Accuracy Takes Power." Decapping is a technique whereby the plastic or ceramic packaging surrounding a mass-produced integrated circuit is delicately removed to reveal the wiring layout inside. This allows the operations of the chip to be mapped and thus recreated.

38. Byuu, "FPGAs Aren't Magic."

39. Any purely analog operations would, of course, definitionally represent a surplus of information over their discrete counterparts. But retro emulation in both its FPGA and software forms targets discrete systems at the level of their discretization—that is, at the point where analog signals are transformed into digital ones. If analog fluctuations in these systems' components influence the digital representations they generate, these effects can themselves be emulated.

40. See Marshall Honorof, "The Console War Is Dead," *Tom's Guide*, January 29, 2022 https://www.tomsguide.com/opinion/console-war-is-dead, accessed July 22, 2022. In this light the entry of Microsoft in the console gaming space would appear to be, in retrospect, a move toward co-opting the specialized and largely Japanese gaming hardware market with the increasingly PC-like Xbox.

41. For many examples see the fascinating https://www.romhacking.net/.

42. Pesce, "Zombie Moore's Law Shows Hardware Is Eating Software." See also *FCP*, 108n6 and note 52 in chapter 1.

43. *FCP*, 101–102 (including table 3.5); Ian Kuon and Jonathan Rose, "Measuring the Gap between FPGAs and ASICs," *IEEE Transactions on Computer-Aided Design of Integrated Circuits and Systems* 26, no. 2 (February 2007): 211–212 (including table IV,

but see also table V). The latter report, while studying later hardware and not methodologically equivalent to *FCP*'s statistics, estimates a 7.1–14-fold advantage in ASIC power consumption over FPGAs (213, including table V, but see also 214 suggesting a potential eighty-seven-fold difference under some circumstances).

44. We can thus see how this term is not only derived from mathematics but has a software parallel in the subroutines of functional programming.

45. Rob Walker with Nancy Tersini, *Silicon Destiny: The Story of Application Specific Integrated Circuits and LSI Logic Corporation* (Milpitas, CA: C.M.C., 1992), 16. On the development of the 4004 out of what was then considered custom circuit design, see Paul E. Ceruzzi, *Computing: A Concise History* (Cambridge, MA: MIT Press, 2012), 101, and Thomas Haigh and Paul E. Ceruzzi, *A New History of Modern Computing* (Cambridge, MA: MIT Press, 2021), 170–171.

46. See, e.g., Waldrop, "More than Moore," 145.

47. Thomas M. Whitney, France Rodé, and Chung C. Tung, "The 'Powerful Pocketful': An Electronic Calculator Challenges the Slide Rule," *Hewlett-Packard Journal* 23, no. 10 (June 1972): 4.

48. Chun observes that the use of the term "program" in computing is similarly derived (Wendy Hui Kyong Chun, "Programmability" in Fuller ed., *Software Studies*, 225; see also Anne Pasek, "From Atoms to Electrons: An Energy History and Future of Computing" in *Digital Energetics*, ed. Anne Pasek et al. (Lüneburg, Germany/Minneapolis: meson/University of Minnesota Press, 2023), 21, and note 94 in chapter 2). Chun also notes more historically that, as we have already seen Near argue in contemporary terms, "as computers became machines . . . programming became functionally equivalent to the process of 'setting up' the ENIAC—the physical act of wiring the machine for a particular problem" (Chun, "Programmability," 225). Elsewhere, she contends that the ENIAC's transition from wired to written programming "dramatically decreased the time necessary for programming while increasing the time necessary for computation," such that, by the early 1960s, an industry-wide "hardware-upgrade cycle was normalized in the name of saving programming time," that is, labor (Wendy Hui Kyong Chun, "On Software, or the Persistence of Visual Knowledge," *Grey Room* 18 (Winter 2004): 28, 31).

49. See, e.g., Karl J. O'Dwyer and David Malone, "Bitcoin Mining and Its Energy Footprint," *Proceedings of the Joint 25th IET Irish Signals & Systems Conference 2014 and 2014 China-Ireland International Conference on Information and Communications Technologies* (June 26–27, 2014), 282. GPUs are discussed at length here in the following chapter in relation to another energy- and computation-heavy process, neural networks.

50. Bitmain Antminer product page, archived at https://web.archive.org/web/20210812080442/https://www.bitmain.com/products/antminer/, accessed August 7, 2024. See also Zane Griffin Talley Cooper, "That Which Escapes: Thinking through Heat in

Proof of Work Systems" in Pasek et al., *Digital Energetics*, 43 and Greenfield, *Radical Technologies*, 332–333n22. For an examination of an older Antminer's capabilities see ibid., 141.

51. Greenfield's (ibid., 115–144) is no exception, taking it to task for its "failure to account for the thermodynamic cost of distributed consensus based on proof-of-work" (140). According to his late-2010s analysis, mining "devices have among the shortest utilization cycles of any commercial information-processing hardware, at significant environmental cost" (333n22). At the same time, however, he argues, "with ASIC hardware having become thoroughly commodified, the mere fact that one is able to marshal large amounts of computational capital is no longer even remotely a guarantee of meaningful return on investment. The only competitive advantage remaining to a prospective miner is energy arbitrage" (138). The continuing obsolescence of older mining hardware suggests, however, that there remains some edge presented by raw computation itself; as I will argue, it is this that the end of Moore's law threatens.

52. New York Environmental Conservation Law §19–0331; Timothy B. Lee, "Ethereum's 'Merge' Is About to Put Every Ether Miner out of Work," *Ars Technica*, August 19, 2022 https://arstechnica.com/tech-policy/2022/08/the-merge-the-biggest-change-in-ethereum-history-explained/, accessed August 20, 2022.

53. Jeffrey West Kirkwood, "From Work to Proof of Work: Meaning and Value After Blockchain," *Critical Inquiry* 48, no. 2 (Winter 2022): 361.

54. See Ralph C. Merkle, "A Digital Signature Based on a Conventional Encryption Function," *Proceedings of the Conference on the Theory and Application of Cryptographic Techniques: Advances in Cryptology—CRYPTO '87* (Lecture Notes in Computer Science vol. 293), ed. Carl Pomerance (Berlin: Springer-Verlag, 1988), 369–378.

55. These processes, and the block header components enumerated previously, are standard parts of the Bitcoin protocol and are also described, e.g., in Kirkwood, "From Work to Proof of Work," 375–377 and Greenfield, *Radical Technologies*, 123–129.

56. Satoshi Nakamoto, "Bitcoin: A Peer-to-Peer Electronic Cash System" (2008) https://bitcoin.org/bitcoin.pdf, accessed July 28, 2022, 1.

57. Kirkwood, "From Work to Proof of Work," 366–367.

58. Ibid., 380.

59. Ibid., 371.

60. Cooper, "That Which Escapes," 45–46.

61. Dead Kennedys, "Kill the Poor," *Fresh Fruit for Rotting Vegetables* (Cherry Red B RED 10, 1980). Aside from the obvious immorality of mass murder, the flaw in

this "thinking," of course, lies in the structural role we have already considered, in which the unemployed serve to regulate wages for those that still receive them.

62. Jonathan Sterne, *MP3: The Meaning of a Format* (Durham, NC: Duke University Press, 2012).

63. For examples of the former, see David Golumbia, *The Politics of Bitcoin: Software as Right-Wing Extremism* (Minneapolis: University of Minnesota Press, 2016); for the latter, see Jonathan Beller, *The World Computer: Derivative Conditions of Racial Capitalism* (Durham, NC: Duke University Press, 2021), e.g., 243–247, and Mark Alizart, *Cryptocommunism*, trans. Robin Mackay (Cambridge, UK: Polity, 2020). Intriguingly, Alizart argues that, as "captive energy," proof-of-work cryptocurrency bridges the hardware–software divide (95).

64. Greenfield, *Radical Technologies*, 140 (citing Charles H. Bennett, "The Thermodynamics of Computation—A Review," *International Journal of Theoretical Physics* 21, no. 12 (December 1982): 905–940).

65. Kirkwood, "From Work to Proof of Work," 366.

66. Carver A. Mead, *Collective Electrodynamics: Quantum Foundations of Electromagnetism* (Cambridge, MA: MIT Press, 2000), *xv*. For more on Mead's inquiry into this topic see the summary of his 2005 remarks in David C. Brock, "Reflections on Moore's Law" in *Understanding Moore's Law: Four Decades of Innovation*, ed. David C. Brock (Philadelphia: Chemical Heritage Press, 2006), 98–99. See also Waldrop, "More than Moore," 146; Pasek, "From Atoms to Electrons," 27–28; and *FCP*, 86.

67. Gordon E. Moore, "Cramming More Components onto Integrated Circuits" (1965), reprinted in *Proceedings of the IEEE* 86, no. 1 (January 1998): 84.

68. Cooper, "That Which Escapes," e.g., 45, 51, 63, 53–54.

69. Ibid., 64.

70. Pasek, "From Atoms to Electrons," 32. See also note 89 in chapter 2. For a further consideration of data centers and their environmental impact see Sean Cubitt, *Finite Media: Environmental Implications of Digital Technologies* (Durham, NC: Duke University Press, 2017), 16–21.

71. Niklaus Wirth, "A Plea for Lean Software," *Computer* 28:2 (February 1995): 64–68. Although colloquially referred to as such, Wirth himself credits this "law" to his colleague Martin Reiser. He, furthermore, disputes its status: "The plague of software explosion is not a 'law of nature.' It is avoidable, and it is the software engineer's task to curtail it" (64, 68); Laura U. Marks, "Calculating and Mitigating Our Streaming Carbon Footprint," in Laura U. Marks, Joseph Clark, Jason Livingston, Denise Oleksijczuk, and Lucas Hilderbrand, "Streaming Media's Environmental Impact," *Media+Environment* 2, no. 1 (2020) https://mediaenviron.org/article/17242-streaming-media-s-environmental-impact, accessed June 20, 2021.

See also W. Stanley Jevons, *The Coal Question: An Inquiry Concerning the Progress of the Nation, and the Probable Exhaustion of Our Coal-mines* (1865), 3rd ed., ed. A. W. Flux (London: Macmillan and Co., 1906) and Peter Frase, *Four Futures: Visions of the World After Capitalism* (London: Verso, 2016), 95–96. The Wikipedia page for the Jevons paradox links to that for Wirth's law as well as "Andy and Bill's law," the industry aphorism about the relationship between hardware and software, with reference, respectively, to the former CEOs of Intel and Microsoft: "what Andy giveth, Bill taketh away" (also quoted in the previously discussed Herb Sutter, "The Free Lunch Is Over: A Fundamental Turn Toward Concurrency in Software" (December 2004/2009) http://www.gotw.ca/publications/concurrency-ddj.htm, accessed April 3, 2017, also published in *Dr. Dobb's Journal* 30, no. 3 (March 2005): 16–22). In relation to Bitcoin mining specifically, see also Greenfield, *Radical Technologies*, 141.

72. Pasek, "From Atoms to Electrons," 35, 31. See also the analysis of data center energy usage and efficiency in *FCP*, 142–145.

73. Pasek, "From Atoms to Electrons," 35.

74. Cooper, "That Which Escapes," 66. What makes it borderline, as I will argue, is the possibility of increased efficiency gains from Moore's law.

75. This is the modern understanding of the maximum efficiency of an engine first established by Carnot (see Nicolas Léonard Sadi Carnot, *Reflections on the Motive Power of Heat* (1824), ed. Robert H. Thurston (1890), 2nd ed. (New York: John Wiley and Sons, 1897), 123–125).

76. Max Planck, *Treatise on Thermodynamics* (1897), 7th ed. (1922), trans. Alexander Ogg (Third English Edition; 1926) (New York: Dover Publications, 1945), 114 (§142). As Planck points out, entropy can also be encountered, independent of energy, as a material configuration (103–104 (§134)). On zero as the lower bound of entropic increase see e.g., 88 (§114), 98–100 (§128–129) and 103 (§133). On reversible processes see also 88 (§115): "Since there exists in nature no process entirely free from friction or heat-conduction, all processes which actually take place in nature, if the second law be correct, are in reality irreversible. Reversible processes form only an ideal limiting case."

77. E.g., Google Cloud Tensor Processing Units, https://cloud.google.com/tpu, accessed August 10, 2024; Cooper, "That Which Escapes," 45.

78. Kirkwood, "From Work to Proof of Work," 364. Similarly, Greenfield argues, Bitcoin "deliberately burns energy—and what's worse, burns it on a calculation without any intrinsic meaning, value or utility" (Greenfield, *Radical Technologies*, 140).

79. But, again, in doing so, they prevent the system from being exploited by attackers who might otherwise sneak fraudulent transactions into computationally trivial block hashes. This is certainly securing *something*, if not meaning.

80. E.g., Shawn James Rosenheim, *The Cryptographic Imagination: Secret Writing from Edgar Poe to the Internet* (Baltimore: Johns Hopkins University Press, 1997).

81. Cooper, "That Which Escapes," 53.

82. Claude E. Shannon, "The Mathematical Theory of Communication" (1948) in Claude E. Shannon and Warren Weaver, *The Mathematical Theory of Communication* (1948/1949) (Urbana: University of Illinois Press, 1998), 31. Shannon's follow-up article explicitly applies these concepts to cryptography (see note 35 in chapter 6).

83. Cooper, "That Which Escapes," 63 (see also the previous page).

84. Andrew Lison, "1968 and the Future of Information," in *The Global Sixties in Sound and Vision: Media, Counterculture, Revolt*, ed. Timothy Scott Brown and Andrew Lison (Basingstoke, UK: Palgrave Macmillan, 2014), 245–274.

85. See Sterne, *MP3* for a more involved discussion of perception and compression.

86. Kittler, "There Is No Software," 85. See also R. N. Horspool and N. Marovac, "An Approach to the Problem of Detranslation of Computer Programs," *The Computer Journal* 23, no. 3 (August 1980): 223–229.

87. Chun, *Programmed Visions*, 24–25 (emphasis in original).

88. Kirkwood, "From Work to Proof of Work," 370.

89. Mike Hazas, Bingsheng Zeng, and Joseph Lindley, "Bitcoin's Energy Use Is Out of Control . . . But Maybe That's the Point," *The Conversation*, December 19, 2017, https://theconversation.com/bitcoins-energy-use-is-out-of-control-but-maybe-thats-the-point-89133, accessed June 20, 2021.

90. See Bitcoin wiki, s.v. "Block Hashing Algorithm," https://en.bitcoin.it/wiki/Block_hashing_algorithm, accessed August 5, 2022.

91. Shannon, "Mathematical Theory of Communication," 56. See also *FCP*, 28 (box 1.3) and Kirkwood, "From Work to Proof of Work," 369.

92. Shannon, "Mathematical Theory of Communication," 62. See also note 84 in chapter 2. Cooper similarly notes that heat, the by-product of this material process, is "largely absent from information theory" ("That Which Escapes," 56).

93. N. Katherine Hayles has argued that, since the injection of noise can paradoxically raise the informational value of a message, information and entropy in the information-theoretical sense "are bound together in a complex dialectic that makes them not so much opposites as complements or supplements to one another" (N. Katherine Hayles, *How We Became Posthuman: Virtual Bodies in Cybernetics, Literature, and Informatics* (Chicago: University of Chicago Press, 1999), 25). See also Kirkwood, "From Work to Proof of Work," 369, Cooper, "That Which Escapes," 56, and note 84 in chapter 2. For an analysis of thermodynamics from a computer-science perspective, including its relationship to information theory, see Paul Cockshott,

Lewis M. Mackenzie, and Greg Michaelson, *Computation and Its Limits* (Oxford, UK: Oxford University Press, 2012), 94–114. They argue that "there turns out to be a deep relationship between computing and thermodynamics, and it is thermodynamics that, on the basis of current understanding, fundamentally limits the power of our computing" (94).

94. In a consideration of Hollywood CGI, Ron Eglash has argued that "like the Marxist observation that 'money is congealed labor,' special effects are congealed computing" (Ron Eglash, "Computing Power," in *Software Studies: A Lexicon*, ed. Matthew Fuller (Cambridge, MA: MIT Press, 2008), 58; quoting a lecture from Donna Haraway). What is novel in proof-of-work cryptocurrencies, however, is that congealed computing can be taken for a kind of money. As Mark Alizart writes, "Because blockchain produces consensus . . . a *bit* can become a *coin*, it can have a rarity that gives it price" (Alizart, *Cryptocommunism*, 63).

95. As Greenfield notes, "even if the authenticity of each [bit]coin is guaranteed . . . the currency is still only as valuable as other people think it is" (Greenfield, *Radical Technologies*, 132; see also Alizart, *Cryptocommunism*, 69). However, for an argument that commodities themselves are and always have been a type of financial instrument, and thus deeply involved with perception and meaning, see Beller, *The World Computer*, e.g. 112–117.

96. Kirkwood, "From Work to Proof of Work," 364; Cooper, "That Which Escapes," 52, 57.

97. Carnot, *Reflections on the Motive Power of Heat*, 42n.

98. *FCP*, 155.

99. Anne Pasek, Cindy Kaiying Lin, Zane Griffin Talley Cooper, and Jordan B. Kinder, "Locating Digital Energetics," in *Digital Energetics*, ed. Pasek et al., 6–7. See also ibid., 2, Cooper, "That Which Escapes," 61 and note 13 in chapter 6.

100. As Anson Rabinbach puts it, "Energy remains entirely indifferent to its social use. Nature may demonstrate no interest in how society employs its gifts, but it is best served by utility or economy of design insofar as the most efficient social use of energy is one . . . with the least amount of waste" (Anson Rabinbach, *The Human Motor: Energy, Fatigue, and the Origins of Modernity* (New York: Basic Books, 1990), 60). We will see in chapter 6 how Carnot's "motive power" became, for Rabinbach via Hermann von Helmholtz, "labor power."

101. Hazas et al., "Bitcoin's Energy Use Is Out of Control."

102. O'Dwyer and Malone, "Bitcoin Mining and Its Energy Footprint." See also Greenfield, *Radical Technologies*, 140.

103. Kirkwood, "From Work to Proof of Work," 370; Cooper, "That Which Escapes," 52.

104. O'Dwyer and Malone, "Bitcoin Mining and Its Energy Footprint," 282.

105. Ibid.

106. Ibid., 283.

107. Kirkwood, "From Work to Proof of Work," 363.

108. As Peter Frase has argued, "while scarce resources have impinged on capitalist development at various points throughout its history, this has repeatedly happened in ways that caught theorists of the system by surprise" (Frase, *Four Futures*, 94).

109. O'Dwyer and Malone, "Bitcoin Mining and Its Energy Footprint," 282. For possible alternatives see Alizart, *Cryptocommunism*, 75.

110. O'Dwyer and Malone, "Bitcoin Mining and Its Energy Footprint," 282.

111. Aoyon Ashraf, "Bitcoin Mining Rig Prices Slump to Near 2-Year Lows Amid Celsius Bankruptcy," *Coindesk*, July 14, 2022 (updated May 11, 2023) https://www.coindesk.com/business/2022/07/14/bitcoin-mining-rig-prices-slump-to-near-2-year-lows-amid-celsius-bankruptcy/, accessed August 7, 2025; Arijit Sarkar, "Global GPU Price Drops to Compensate for Falling Bitcoin Mining Revenue," *Cointelegraph*, July 9, 2022 https://cointelegraph.com/news/global-gpu-price-drops-to-compensate-for-falling-bitcoin-mining-revenue, accessed August 4, 2022. See also note 105 earlier.

112. Kirkwood, "From Work to Proof of Work," 368. See also note 51 earlier. Compare this hypothetical to one in which speed continues improving but efficiency does not, raising the proportional cost of energy vs. hardware (*FCP*, 142–143).

113. Ibid., 97.

114. As Jonathan Beller notes, Bitcoin's "value is tied to the vitality of the network" (Beller, *The World Computer*, 244).

115. For an in-depth analysis of this weakness in the protocol see Greenfield, *Radical Technologies*, 139–140.

116. Hazas et al., "Bitcoin's Energy Use Is Out of Control."

117. Lee, "Ethereum's 'Merge.'"

118. Even without considering an end to Moore's law, Cubitt, citing Nicholas Carr, notes that such expenditure concerns motivate data center users in general, "rather than sustainability in its own right" (Cubitt, *Finite Media*, 20).

119. Andrew Lison, "Hardware Standardization and State-Socialist Piracy: The Global Reach of the Zilog Z80," *IEEE Annals of the History of Computing* 46, no. 2 (April–June 2024): 49.

120. E.g., Chun, *Programmed Visions*, 156: "The logic of computers as logos stems from the disciplining, the axiomatizing, of hardware," a process she relates to the

development of neural networks, and as such will be discussed further in the following chapter. Here, however, see also 26–27.

121. See, e.g., Mél Hogan, "Data Flows and Water Woes: The Utah Data Center," *Big Data & Society* 2, no. 2 (July–December 2015): 1–12.

122. Hazas et al., "Bitcoin's Energy Use Is Out of Control."

123. *FCP*, 144–145.

124. Cubitt, citing noted power usage scholar Jonathan Koomey, makes a similar argument regarding data center usage in general vs. the economy overall (Cubitt, *Finite Media*, 21).

125. Efforts at such analysis by Bitcoin proponents taking into account much (but not all) of the above conclude, perhaps unsurprisingly, that it is *more* efficient, e.g., Hass McCook, "Bitcoin Emits Less than 5% of the Legacy Financial Sector's Carbon Emissions," *Bitcoin Magazine*, May 24, 2021, https://bitcoinmagazine.com/business/bitcoin-vs-financial-sector-energy-use, accessed January 15, 2025. Similarly, see Alizart, *Cryptocommunism*, 129n2 for a favorable comparison, circa 2019, of Bitcoin with cooling systems globally. Hazas et al., "Bitcoin's Energy Use Is Out of Control" cites earlier work by McCook to argue that "computers required to 'mine' digital coins actually use more energy, and ultimately create more carbon emissions, per unit of value created than the machines and resources used to print and mint traditional currency," but admits this does not take into account their ability to do so without bank or state backing. Without endorsing either side of this argument, there is clearly room for further dialogue and analysis.

126. Benedict Anderson, *Imagined Communities: Reflections on the Origin and Spread of Nationalism* (1983) Revised Edition (London: Verso, 2006).

127. Bertolt Brecht, *The Threepenny Opera* (1928), trans. Desmond Vesey and Eric Bentley (1960) (New York: Grove Press, 1964), 92 (Act 3, Scene 3).

128. To his credit, Kirkwood does acknowledge the present-day exploitation of the Congo as one of the ways "that the computational economy still relies on, and in many ways exacerbates, the worst elements of global industrial extraction" (Kirkwood, "From Work to Proof of Work," 368n20), but he makes little attempt to put these two forms of valorization into relation.

129. Cubitt, *Finite Media*, 24. For Cubitt, this distinction is aligned with one between excess consumption and excess production (7). However, as I will argue in chapter 5, thanks to its thus far relentless pace of innovation the semiconductor industry has a different relationship to overproduction than typical commodity manufacturing.

130. Pasek, "From Atoms to Electrons," 37.

131. Carnot, *Reflections on the Motive Power of Heat*, 126.

Chapter 4

1. Aside from the Cerebras chips and Google's Tensor Processing Units mentioned previously, hardware specifically designed for AI includes Apple's A12 Bionic SoC, NVIDIA's EGX platform ranging from single-board computers to servers, Intel's Gaudi processors, and Microsoft's Project Brainwave FPGA effort. AI hype in the press is widespread, e.g., the series "How AI Is Changing Everything" in *Business Insider*, https://www.businessinsider.com/how-ai-is-changing-everything-2019-7, accessed February 9, 2020. Electronic musician and composer Holly Herndon's album *Proto* (4AD 4AD0140LP, 2019) is advertised as having been made with the assistance of an AI dubbed "Spawn." The intersection of AI and ethics has moved from an obscure subfield to a major endeavor for many universities with academic centers and initiatives such as the Stanford Institute for Human-Centered Artificial Intelligence, Berkeley's Center for Human-Compatible AI, and Harvard's Berkman Klein Center's Ethics and Governance of AI Initiative, as well as a focus of independent institutes such as Data & Society and the AI Ethics Lab and publications such as Patrick Lin, Keith Abney, and Ryan Jenkins, eds., *Robot Ethics 2.0: From Autonomous Cars to Artificial Intelligence* (Oxford, UK: Oxford University Press, 2017), Mark Coeckelbergh, *AI Ethics* (Cambridge, MA: MIT Press, 2020), and Markus D. Dubber, Frank Pasquale, and Sunit Das, eds., *The Oxford Handbook of Ethics of AI* (Oxford, UK: Oxford University Press, 2020).

2. Ben Buchanan and Andrew Imbrie, *The New Fire: War, Peace, and Democracy in the Age of AI* (Cambridge, MA: MIT Press, 2022); Shoshana Zuboff, *The Age of Surveillance Capitalism: The Fight for a Human Future at the New Frontier of Power* (New York: Public Affairs, 2019), e.g. 65; Ruha Benjamin, *Race After Technology: Abolitionist Tools for the New Jim Code* (Cambridge, UK: Polity, 2019). All these aspects are touched upon in the discussion of machine learning and AI in Adam Greenfield, *Radical Technologies: The Design of Everyday Life* (London: Verso, 2017), 209–271.

3. See for example Eliezer Yudkowsky, "The AI Alignment Problem: Why It's Hard, and Where to Start," Machine Intelligence Research Institute white paper/Stanford University Symbolic Systems Distinguished Speaker Lecture, May 6, 2016, available online at https://intelligence.org/files/AlignmentHardStart.pdf, accessed December 2, 2020, and Iason Gabriel, "Artificial Intelligence, Values, and Alignment," *Minds and Machines* 30, no. 3 (September 2020): 411–437.

4. E.g., Justin Joque, *Revolutionary Mathematics: Artificial Intelligence, Statistics and the Logic of Capitalism* (London: Verso, 2022).

5. Luciana Parisi, "Automated Cognition and Capital," in *The Psychopathologies of Cognitive Capitalism: Part Three*, ed. Warren Neidich (Berlin: Archive Books, 2017), 71.

6. Will Knight, "The Dark Secret at the Heart of AI," *MIT Technology Review*, April 11, 2017, https://www.technologyreview.com/2017/04/11/5113/the-dark-secret-at-the-heart-of-ai/, accessed May 18, 2020. Knight's article offers an in-depth examination

of this aspect, one upon whose particulars the present study will necessarily also touch. Naturally, and as Knight describes, efforts are ongoing to address this concern, but they often build on top of this fundamental uncertainty, encouraging post hoc "explanations" of results at higher, rather than lower, levels of operation.

7. Wendy Hui Kyong Chun, *Programmed Visions: Software and Memory* (Cambridge, MA: MIT Press, 2010).

8. The same phrase appears in the novel upon which the film was based (D. F. Jones, *Colossus* (1966) (New York: G. P. Putnam's Sons, 1967), 77).

9. With reference to Hegel's evocation of the foundational brilliance of Plato in G. W. F. Hegel, *The Philosophy of Right* (1821) trans. S. W. Doyle (1896) (Mineola, NY: Dover, 2005), *xix*. For this process of unification see G. W. F. Hegel, *Phenomenology of Spirit* (1807) trans. A. V. Miller (Oxford, UK: Oxford University Press, 1979). Kittler, once again, is prescient, writing in the early 1990s of "the fairy tale that software was being developed which would become increasingly gentle, user-friendly, spiritual, and intelligent: one day in the not-so-distant future it would effectively amount to German Idealism—that is, software would become human" (Friedrich A. Kittler, "Protected Mode" (1991) in *The Truth of the Technological World*, trans. Erik Butler (Stanford, CA: Stanford University Press, 2013), 209). Computer scientists, however, would seem to be anti-Hegelians on this point. Addressing the difficulty of comprehending CNN AI, Knight quotes a researcher musing that "it might just be part of the nature of intelligence that only part of it is exposed to rational explanation. Some of it is just instinctual, or subconscious, or inscrutable" (Jeff Clune quoted in Knight, "The Dark Secret"). Similarly, W. Daniel Hillis observes of "evolved" programs, developed through an artificial process of successful competition and subsequent procreation rather than direct coding, "if evolution can produce something as simple as a sorting program which is fundamentally incomprehensible—it does not bode well for our prospects of ever understanding the human brain" (W. Daniel Hillis, *The Pattern on the Stone: The Simple Ideas That Make Computers Work* (1998), 2nd ed. (New York: Basic Books, 2015), 147 (see also 141 and Greenfield, *Radical Technologies*, 253)). Finally, for a philosophical refutation of the possibility of artificial general intelligence, on the basis that, irrespective of improvements to computing power, a constitutive inability to mathematically model sufficiently complicated systems, including the brain, prevents its realization, see Jobst Landgrebe and Barry Smith, *Why Machines Will Never Rule the World: Artificial Intelligence Without Fear* (New York and London: Routledge, 2023).

10. This description is synthesized from the discussions in the texts cited herein. For similar brief overviews see Andrew Goffey, "Intelligence," in *Software Studies: A Lexicon*, ed. Matthew Fuller (Cambridge, MA: MIT Press, 2008), 138 and Greenfield, *Radical Technologies*, 215–218. For a detailed description of the overall process of convolutional neural network training, see the key paper by Yann LeCun, Léon Bottou, Yoshua Bengio, and Patrick Haffner, "Gradient-based Learning Applied to Document Recognition," *Proceedings of the IEEE* 86, no. 11

(November 1998): 2278–2324. On back-propagation see also Knight, "The Dark Secret." For its development see David E. Rumelhart, Geoffrey E. Hinton, and Ronald J. Williams, "Learning Representations by Back-propagating Errors," *Nature* 323 (October 9, 1986): 533–536. For the roots of this procedure in control theory see LeCun et al., "Gradient-based Learning," 2281. Earlier cybernetic models of neurons treated their configurations as static (Chun, *Programmed Visions*, 154).

11. J. L. McClelland, D. E. Rumelhart, and G. E. Hinton, "The Appeal of Parallel Distributed Processing," (1986) in George F. Luger ed., *Computation & Intelligence: Collected Readings* (Menlo Park, CA/Cambridge, MA: American Association for Artificial Intelligence/MIT Press, 1995), 313. Hinton was awarded the 2024 Nobel Prize in Physics in part due to the connections he drew between physics and AI.

12. Jiuxiang Gu et al., "Recent Advances in Convolutional Neural Networks," *Pattern Recognition* 77 (2018): 371. See also Knight, "The Dark Secret" and Greenfield, *Radical Technologies*, e.g., 244 and 252.

13. On feature identification see ibid., 215. For a visual example of what some intermediate or "hidden" feature maps are trained to detect see LeCun et al., "Gradient-based Learning," 2295 (figure 13). Knight, "The Dark Secret," similarly features artworks by Adam Ferriss taking these intermediate layers as their basis.

14. Alex Krizhevsky, Ilya Sutskever, and Geoffrey E. Hinton, "ImageNet Classification with Deep Convolutional Neural Networks," (2012) in *Communications of the ACM* 60, no. 6 (June 2017): 84. See also LeCun et al., "Gradient-based Learning," 2278–2279, 2294, 2995 (figure 13), and 2305 and Greenfield, *Radical Technologies*, 215–216.

15. Already an issue with self-driving cars of any sort (see Corinne Iozzio, "Who's Responsible When a Car Controls the Wheel?" in *Scientific American* 314, no. 5 (May 2016): 12–13), this concern is exacerbated by the specificities of AI (see, e.g., Knight, "The Dark Secret," Ron Schmelzer, "What Happens When Self-Driving Cars Kill People?" in *Forbes*, September 26, 2019 https://www.forbes.com/sites/cognitiveworld/2019/09/26/what-happens-with-self-driving-cars-kill-people/, accessed February 9, 2020) and has been popularized in films, e.g., Werner Herzog's *Lo and Behold: Reveries of the Connected World*. See also Greenfield, *Radical Technologies*, 253 and, on self-driving cars more broadly, passim, e.g., 222–226.

16. Knight, "The Dark Secret."

17. Chun, *Programmed Visions*, 102–103 and 140–141, 62; see also John von Neumann, "First Draft of a Report on the EDVAC" (United States Army Ordnance Department Contract W-670-ORD-4926), Moore School of Electrical Engineering, University of Pennsylvania (June 30, 1945).

18. Chun, *Programmed Visions*, 155–162, 61–62.

19. Ibid., 9.

20. See Kyoung-Su Oh and Keechul Jung, "GPU Implementation of Neural Networks," *Pattern Recognition* 37 (2004): 1311–1314.

21. Krizhevsky et al., "ImageNet Classification." See also Tim Hwang, "Computational Power and the Social Impact of Artificial Intelligence," March 23, 2018, https://ssrn.com/abstract=3147971, 5–6, accessed February 5, 2022).

22. Tekla S. Perry, "Move Over Moore's Law: Make Way for Huang's Law," *IEEE Spectrum* April 2, 2018, https://spectrum.ieee.org/move-over-moores-law-make-way-for-huangs-law, accessed August 8, 2022. As Jacob Gaboury notes about Intel engineer Raja Koduri's similar "law" describing the rate of change in computing architecture, "whether this law is accurate or logical is questionable, as its goal seems to be to propose the inevitability of a coming age of 'Discrete AI' that will replace GPU computing rather than serving as an accurate historical descriptor" (Jacob Gaboury, *Image Objects: An Archaeology of Computer Graphics* (Cambridge, MA: MIT Press, 2021), 256n16; see also ibid., 199).

23. Joseph Weizenbaum, *Computer Power and Human Reason: From Judgment to Calculation* (San Francisco: W. H. Friedman and Co., 1976), 3–4.

24. Ibid., 7.

25. McClelland et al., "The Appeal of Parallel Distributed Processing," 340.

26. Ibid. Hwang makes a similar point about NN research in the 1950s and '60s (Hwang, "Computational Power and the Social Impact of Artificial Intelligence," 4).

27. See e.g. Hubert L. Dreyfus, *What Computers* Still *Can't Do* (1972/1979/1992), (Cambridge, MA: MIT Press, 1992), *xxi–xxii*, McClelland et al., "The Appeal of Parallel Distributed Processing," 313, and Goffey, "Intelligence," 137–138.

28. On the former example see Alan M. Turing, "Computing Machinery and Intelligence" in *Mind* 59:236 (October 1950): 433–460. Hwang characterizes the latter example in terms of modern AI applications' "latency" (Hwang, "Computational Power and the Social Impact of Artificial Intelligence," 9). The logical endpoint of this move is so-called real-time computing, which, as Chun points out, can never strictly speaking operate as such (Chun, *Programmed Visions*, 68).

29. Flusser argues that computers invert the Platonic order by proceeding from, rather than arriving at, ideal forms (Vilém Flusser, "Pots" in *The Shape of Things: A Philosophy of Design* (1993), trans. Anthony Mathews (London: Reaktion Books, 1999), 99–103). From the present perspective, this outlook is derived from classical programming paradigms rather than neural models of computation, which, despite being abstract, do not impose their form directly upon their output but rather derive the latter from an encounter between input and a formal model of brain-like functionality. Dreyfus, although as we shall see particularly skeptical of classical programming methods for AI, would seem to agree that their paradigms are essentially Platonic. Digital computers, he writes, "might have remained overgrown adding

machines, had not Plato's vision, refined by two thousand years of metaphysics, found in them its fulfillment" (Dreyfus, *What Computers Still Can't Do*, 72).

30. John H. Holland, "Escaping Brittleness: The Possibilities of General-Purpose Learning Algorithms Applied to Parallel Rule-Based Systems" (1986) in Luger ed., *Computation & Intelligence*, 279. See also Goffey, "Intelligence," 138. Most works examined here identify the 1980s as a conceptual—but crucially, not material—turning point for NN AI.

31. Holland, "Escaping Brittleness," 278.

32. McClelland et al., "The Appeal of Parallel Distributed Processing," 338.

33. Ibid., 306.

34. Dreyfus, *What Computers Still Can't Do, ix, xiv*. Dreyfus takes the term GOFAI from John Haugeland. The narrative of this split and the eventual triumph of the latter approach is relatively canonical and can also be found in e.g. Knight, "The Dark Secret" and Greenfield, *Radical Technologies*, 213–214.

35. Dreyfus, *What Computers Still Can't Do, xiv*.

36. Catherine Malabou, *What Should We Do with Our Brain?* (2004), trans. Sebastian Rand (New York: Fordham University Press, 2008), 53. See also Marc Jeannerod's foreword (*xiii*). Malabou suggestively connects the development of what she calls *"neuronal ideology"* not simply with the end of state socialism but the socio-political upheavals of the 1960s as well (11 (emphasis in original), 41).

37. Ibid., 65; Chun, *Programmed Visions* 102–103.

38. Holland, "Escaping Brittleness," 277, 294. Holland's narrative does, however, echo Chun's characterization of Charles Davenport's approach to eugenics, in which "humans are both evaluated in terms of their economic costs to society and viewed as a form of capital"; drawing on Michel Foucault in both instances, she elsewhere writes that "typical neoliberal fashion . . . evaluates all activities in terms of cost-benefit analysis" (Chun, *Programmed Visions*, 123 (see also 125), 84. See also Michel Foucault, *The Birth of Biopolitics: Lectures at the Collège de France, 1978–1979*, ed. Michel Senellart (2004), trans. Graham Burchell (Basingstoke, UK: Palgrave Macmillan, 2008)). For a more recent consideration of these intersections see Orit Halpern, "The Future Will Not Be Calculated: Neural Nets, Neoliberalism, and Reactionary Politics," *Critical Inquiry* 48, no. 2 (Winter 2022): 334–359. I consider both Foucault and eugenics in relation to the birth of the transistor in the following chapter.

39. Malabou, *What Should We Do with Our Brain?*, 46.

40. Ibid., 12. See also Chun, *Programmed Visions*, 63 and 130–131.

41. McClelland et al., "The Appeal of Parallel Distributed Processing," 313.

42. Dreyfus, *What Computers Still Can't Do, xxxv.*

43. Ibid., *l*n46. Or, as Hillis puts it, with neural networks, "the catch is the assumption that there exists some correct pattern of weights that will accomplish the task" (Hillis, *The Pattern on the Stone*, 131). See also Goffey, "Intelligence," 138. Malabou argues that a similar situation holds true for human brains (Malabou, *What Should We Do with Our Brain?*, 6).

44. Dreyfus, *What Computers* Still *Can't Do, xxxvi* (see also Greenfield, *Radical Technologies*, 218). At a more technical level, LeCun et al. describe a similar issue in the training phase they call "the collapse problem": "If the recognizer is a simple NN . . . the minimum of the loss function is attained, not when the recognizer always gives the right answer, but when it ignores the input and sets its output to a constant vector with small values for all the components." (LeCun et al., "Gradient-based Learning," 2300, but see also 2286).

45. Dreyfus, *What Computers Still Can't Do, xxxiii.*

46. Ibid., *xlv.*

47. See Gene M. Amdahl, "Validity of the Single Processor Approach to Achieving Large Scale Computing Capabilities," *American Federation of Information Processing Societies Conference Proceedings*, vol. 30 (Spring 1967), 483–485 and The National Research Council Committee on Sustaining Growth in Computing Performance, *The Future of Computing Performance: Game Over or Next Level?*, eds. Samuel H. Fuller and Lynette I. Millett (Washington, DC: National Academies Press, 2011), e.g., 64, box 2.2 (65), and box 5.1 (134) (hereafter referred to as *FCP*). For a lay-accessible treatment, along with the argument that the parallel architecture of the brain offers a way around this limit, see Hillis, *The Pattern on the Stone*, 112–120. Amdahl's law presents complications for Benjamin Bratton's suggestion "that there is something like a Moore's law of exponential computing speed . . . for the [network] interfaces and the landscapes of gateways that they power," one whose "exponential arc of interfacial acceleration might measure the total aggregate switches within a landscape of interfaces" (Benjamin H. Bratton, *The Stack: On Software and Sovereignty* (Cambridge, MA: MIT Press, 2015, 232).

48. See, e.g., Hillis, *The Pattern on the Stone*, 127.

49. Amdahl, "Validity of the Single Processor Approach," 484f.

50. Bernhard E. Boser et al., "An Analog Neural Network Processor with Programmable Topology," in *IEEE Journal of Solid-State Circuits* 26, no. 12 (December 1991): 2017.

51. Ibid., 2023.

52. Ibid., 2024.

53. LeCun et al., "Gradient-based Learning," 2293.

54. Boser et al., "Analog Neural Network Processor," 2021–2022; *Intel Microprocessor Quick Reference Guide* (2008), https://www.intel.com/pressroom/kits/quickreffam.htm, accessed February 14, 2020.

55. Ibid. In actuality LeNet-5 appears to have been trained on a 200MHz MIPS processor (LeCun et al., "Gradient-based Learning," 2293), ten times the clock speed of the previous paper's chip (Boser et al., "Analog Neural Network Processor," 2023).

56. As Mark Pesce writes, historically, "the cheap gains of ever-faster CPUs versus the hard work of designing and debugging silicon circuitry meant only the most important or time-critical tasks migrated into silicon" (Mark Pesce, "Zombie Moore's Law Shows Hardware Is Eating Software," *The Register*, September 22, 2016, https://www.theregister.com/2016/09/22/the_evolution_of_moores_law_suggests_hardware_is_eating_software/ accessed July 27, 2025). See also *FCP*, 54.

57. LeCun et al., "Gradient-based Learning," 2279, 2307.

58. See, e.g., Oh and Jung, "GPU Implementation of Neural Networks" and Hwang, "Computational Power and the Social Impact of Artificial Intelligence," 10–11. Although not addressing NNs directly, on clock speed, instruction sets, parallel processing, and GPUs see also *FCP*, e.g., 16, Box 2.4 (72–73), and 138. Despite touching on individual techniques and applications like machine learning, matrix multiplication and convolution, and image analysis (e.g., 51, 112–113, and 133, respectively), this 2011 report's claim that "we do not have new software approaches that can exploit the innovative [multiprocessing hardware] architectures" (*viii*) demonstrates the rapidity with which the AI paradigm shift has arisen.

59. This description of the overall operation of CNNs is synthesized from many of the works cited in this chapter but in particular Knight, "The Dark Secret" and, more significantly and as with the description of training above, LeCun et al., "Gradient-based Learning," e.g., 2283–2284.

60. Oh and Jung, "GPU Implementation of Neural Networks" describes these internal features of GPUs and how they can be used for CNNs. Their research is based on a single ATI Radeon 9700 Pro card. Krizhevsky et al., "ImageNet Classification" makes use of newer parallelization features between two NVIDIA GTX 580 GPUs (86). In his "Huang's law" speech, Huang touted the improvements between this last process and NVIDIA's then cutting-edge DGX-2 AI-focused computer (Perry, "Move Over Moore's Law"). This design contains sixteen Tesla V100 GPUs (NVIDIA, DGX-2 datasheet (2019), https://www.nvidia.com/content/dam/en-zz/Solutions/Data-Center/dgx-1/dgx-2-datasheet-us-nvidia-955420-r2-web-new.pdf, accessed August 8, 2022). Stronger systems continue to be made.

61. Herb Sutter, "The Free Lunch Is Over: A Fundamental Turn Toward Concurrency in Software" (December 2004/2009) http://www.gotw.ca/publications/concurrency-ddj.htm, accessed April 3, 2017, also published in *Dr. Dobb's Journal* 30, no. 3 (March 2005): 16–22.

62. *FCP*, 132.

63. See, e.g., Hwang, "Computational Power and the Social Impact of Artificial Intelligence," 11.

64. Ibid., 3. Hwang covers many of the post-2004 developments discussed here and previously, including his own analysis of FPGAs and ASICs (11–13), but his analysis differs substantially, as when he sees hardware and software as diverging rather than synthesizing (8); more broadly, his attention focuses on the ethical, industrial, geopolitical, and—as we shall see—governance implications of AI, rather than barriers to Moore's law's perpetuation. Where we concur is on the fundamentally computational and material nature of AI (2, 15). Further, as I have already made clear, this routine and frequently unexamined—which is to say, ideological—appeal to unspecified increases in computational power is in fact endemic not only to AI but nearly every aspect of digital hype and boosterism. Indeed, I would suggest that many of Hwang's points—for example that "Machine learning . . . is inextricably tied to computational power—the tangible material of chips and semiconductors that the algorithms of machine intelligence operate on" (2) and his examination of the way certain ML developments could facilitate the replacement of labor with computation (40–42)—are best understood as statements about the significance of computation and computing power in general, rather than the AI-specific assertions he is directly making.

65. LeCun et al., "Gradient-based Learning," 2279; Krizhevsky et al., "ImageNet Classification," 90 (in an epilogue to the original 2012 paper). Knight, "The Dark Secret" also underscores the first two factors. Hwang makes a similar point about the elision of computing power with reference to a 2016 White House National Science and Technology Council paper (Hwang, "Computational Power and the Social Impact of Artificial Intelligence," 7). Large datasets increase the applicability of parallel processing (*FCP*, e.g., 62, 136).

66. Rui-Jie Zhu et al., "Scalable MatMul-free Language Modeling," *arXiv* preprint 2406.02528 [cs.CL] https://doi.org/10.48550/arXiv.2406.02528, accessed June 16, 2024.

67. Krizhevsky et al., "ImageNet Classification," 85. Adam Greenfield, similarly, argues that "given only sufficient processing power, . . . sufficiently well-trained feature-extraction algorithms, and access to a sufficiently large corpus of example works, abstracting" human creativity "is not much more than trivial" (Greenfield, *Radical Technologies*, 261–262).

68. Hwang, "Computational Power and the Social Impact of Artificial Intelligence," 16.

69. Sutter, "The Free Lunch Is Over."

70. See, e.g., *FCP*, 73 (Box 2.4)

71. Friedrich Kittler, "There Is No Software," *Stanford Literature Review* 9.1 (Spring 1992): 90, 89.

72. See note 33 in chapter 1.

73. Michael Conrad, "The Price of Programmability," in *The Universal Turing Machine: A Half-Century Survey* (1988), ed. Rolf Herken, 2nd ed. (Vienna: Springer-Verlag, 1994/1995), 275 (emphasis in original).

74. Ibid., 276–278.

75. Boser et al., "Analog Neural Network Processor," 2018, 2023.

76. See Malabou, *What Should We Do with Our Brain?*, 36.

77. Conrad, "The Price of Programmability," 279.

78. See, e.g., Catherine Liu, "The Humanities and Artisanal Education: Technocrats and Artisans in the Contemporary University," *College Literature* 42, no. 2 (Spring 2015): 281, Gary Hall, *The Uberfication of the University* (Minneapolis; University of Minnesota Press, 2016), especially 31–34, and Malabou's discussion of "employability" more generally (Malabou, *What Should We Do with Our Brain?*, 45–46).

79. David F. Noble, *Digital Diploma Mills: The Automation of Higher Education* (New York: Monthly Review Press, 2001), 26. See also Liu, "The Humanities and Artisanal Education," 281.

80. Ibid., 283.

81. Ibid., 286.

82. Hall, *The Uberfication of the University*, 31–32.

83. Noble, *Digital Diploma Mills*, 29.

84. Ibid., 32. This is the main focus of Hall, *The Uberfication of the University*.

85. Noble, *Digital Diploma Mills*, 33.

86. Hall, *The Uberfication of the University*, 37.

87. Malabou, *What Should We Do with Our Brain?*, 37. "Weak AI" is a term traceable back to philosopher John Searle but which Malabou takes, via Jean-François Dortier, to describe the shift toward "AI" no longer geared toward reconstructing human thought processes but rather accomplishing individual tasks previously considered exclusive to humans (89n12).

88. CaptRobau's Place https://captrobau.blogspot.com/; Timothy B. Lee, "Someone Used Neural Networks to Upscale a Famous 1896 Video to 4k Quality (Updated)," *Ars Technica*, February 4, 2020 (updated February 5), https://arstechnica.com/science/2020/02/someone-used-neural-networks-to-upscale-a-famous-1896-video-to-4k-quality/ (URLs accessed February 19, 2020).

89. Ilia Shumailov et al., "AI Models Collapse When Trained on Recursively Generated Data," *Nature* 631 (2024): 755–759.

90. For a critical analysis of GPT-2 and its successor, GPT-3, see Landgrebe and Smith, *Why Machines Will Never Rule the World*, 240–241.

91. Malabou, *What Should We Do with Our Brain?*, 35, 59, 60. Chun, similarly, argues that, while it abstracted away from real brains, in the cybernetic model of memory "signals bridge mind and brain because they have a double nature; they are both physical events and symbolic values" (Chun, *Programmed Visions*, 156).

92. Malabou, *What Should We Do with Our Brain?*, 62–63.

93. See Claude E. Shannon and Warren Weaver, *The Mathematical Theory of Communication* (1948/1949) (Urbana: University of Illinois Press, 1998).

94. See, e.g., Goffey, "Intelligence," 138.

95. Slavoj Žižek, *The Sublime Object of Ideology* (1989), new edition (New York: Verso, 2009), 29–30.

96. Chun, *Programmed Visions*, 59, 67, 71–75, 52–53.

97. Ibid., 130. Chun links this shift to the massive amount of information produced by the Human Genome Project—that is, to the growth of collected datasets—which she argues led to an emphasis on both interconnection in biology and the modeling, rather than the interpretation, of data flows that forms the bedrock of CNN AI (129–130). However, Conrad's work, not to mention the "connectionist" school of cognitive science identified by Dreyfus, predates this development.

98. E.g. Ibid., 36, 45; 143–144 (noting that early analog computers, unlike their digital counterparts, could conduct concurrent operations); 152–157, 164.

99. This remains a problem even for Malabou's consideration of neuroscience directly, where she moves from a "universal" biological foundation for human consciousness to the particularities of individual subject formation but tantalizingly leaves the question of a higher-order synthetic recuperation of this contradiction, gestured toward in her titular question, for the future (see, e.g., Malabou, *What Should We Do with Our Brain?*, 62).

100. Giorgio Agamben, *What Is Real?* (2016) trans. Lorenzo Chiesa (Stanford, CA: Stanford University Press, 2018), 42. Nonetheless, as ever when we trace the deconstructive tradition far back enough, we eventually uncover the presence of Heidegger: "Physics, in all its retreating from . . . representation . . . will never be able to renounce this one thing: that nature reports itself in some way or other that is identifiable through calculation and that it remains orderable as a system of information . . . It seems as though causality is shrinking into a reporting . . . To this

shrinking would correspond the process of growing resignation that Heisenberg's lecture depicts in so impressive a manner" (Martin Heidegger, "The Question Concerning Technology" (1954) in *The Question Concerning Technology and Other Essays* (1977) trans. William Lovett (New York: Harper Perennial Modern Thought, 2013), 23).

101. Ettore Majorana, "The Value of Statistical Laws in Physics and Social Sciences" (1942) trans. Rosario Nunzio Mantegna (2005) in Agamben, *What Is Real?*, 65. "According to Weil, renouncing necessity and determinism in the name of probability, quantum mechanics has purely and simply renounced science" (Agamben, *What Is Real?*, 23). For a broader account of the rise of statistical thinking and its relation to physics see Mary Ann Doane, *The Emergence of Cinematic Time: Modernity, Contingency, The Archive* (Cambridge, MA: Harvard University Press, 2002), 108–139, especially 122–130.

102. E.g., Norbert Wiener, *Cybernetics; or, Control and Communication in the Animal and Machine* (1948), 2nd ed. (1961) (Cambridge, MA: MIT Press, 2019), 18; Norbert Wiener, *The Human Use of Human Beings: Cybernetics and Society* (1950) Revised Edition (1954) (Cambridge: Da Capo Press, 1988), 15–16, 95–96, and 151–152; J. Clerk Maxwell, "On Governors," in *Proceedings of the Royal Society of London* 16 (1867–1868): 270–283; Shannon and Weaver, *The Mathematical Theory of Communication.* This genealogy is discussed further in chapter 6.

103. Malabou, *What Should We Do with Our Brain?*, 34. On neural homeostasis, see, e.g., 74–75. On plasticity, a core and explicitly dialectical element of Malabou's philosophy, see, e.g., 4–31, Chun, *Programmed Visions*, 131, and, as a quality of AI specifically, Goffey, "Intelligence," 138.

104. Friedrich Kittler, "Code (or, How You Can Write Something Differently)" (2003) trans. Tom Morrison with Florian Cramer in Fuller ed., *Software Studies*, 46. Or, as Malabou puts it regarding contemporary neuroscience, "Not to interpret is still to interpret. By wishing not to construct a hermeneutic schema capable of explaining, at least provisionally, the relations between the neuronal and the mental . . . one exposes oneself, whether one recognizes it or not, to ideological drift" (Malabou, *What Should We Do with Our Brain?*, 64–65).

105. McClelland et al., "The Appeal of Parallel Distributed Processing," 313.

106. Elizabeth Wilson, *Affect and Artificial Intelligence* (Seattle: University of Washington Press, 2010), 4.

107. Writing more broadly, Hall asks if tech "companies and their products *are* saving people's time or whether they are actually enabling them to work even more" (Hall, *The Uberfication of the University*, 10).

108. Goffey, "Intelligence," 133, 134.

109. Mechanical Turk workers were used to classify the data with which the winning 2012 ImageNet CNN was trained (see Krizhevsky et al., "ImageNet Classification," 85, and Hwang, "Computational Power and the Social Impact of Artificial Intelligence," 5).

110. Ibid., 47.

111. Alain Badiou, *Ethics: An Essay on the Understanding of Evil* (1998) trans. Peter Hallward (London: Verso, 2001), 9.

112. Malabou, *What Should We Do with Our Brain?*, 13.

113. Hillis, *The Pattern on the Stone*, 118–119. See also Landgrebe and Smith, *Why Machines Will Never Rule the World*, 139.

114. For Jameson, "History is . . . the experience of Necessity . . . the inexorable form of events . . . History is what hurts, it is what refuses desire and sets inexorable limits to individual as well as collective praxis" (Fredric Jameson, *The Political Unconscious: Narrative as a Socially Symbolic Act* (Ithaca, NY: Cornell University Press, 1981), 102). The question of necessity is also central to Weil's critique of quantum physics, on which Agamben builds: "Some splendid verses of Lucretius suffice to give us a sense of what is purifying in the sight and experience of necessity; well-endured affliction is a purification of this kind, as is classical science also a purification, if one makes good use of it, for it tries to read through all appearances this inexorable necessity that makes of the world a world where we do not count, a world where we work, a world indifferent to desire, aspirations, and the good; it is the study of the sun that shines indifferently on the just and the unjust" (Weil quoted in Agamben, *What Is Real?*, 22–23; see also note 101 earlier).

115. Walter Benjamin, "Theses on the Philosophy of History" (1940) in *Illuminations* (1955) trans. Harry Zohn, ed. Hannah Arendt (New York: Schocken Books, 1968/2007), 253. Scholars have long noted the connection between this text, AI, and Amazon's service, e.g., Mark Sample, "The Amazing Amazon Mechanical Turk," *@samplereality*, July 31, 2006, https://samplereality.com/2006/07/31/the-amazing-amazon-mechanical-turk/, accessed June 28, 2021.

116. Walter Benjamin, *The Arcades Project*, trans. Howard Eiland and Kevin McLaughlin (Cambridge, MA: Belknap/Harvard University Press, 1999), 456–488. On the translation of "convolute" see ibid., *xiv*.

117. "You would not evoke the messianic in a genuinely revolutionary period . . . it is a unique variety of the species hope that . . . flourishes only in a time of absolute hopelessness, a period like the Second Empire, or the between the war years, or the 1980s and 90s, when radical change seems unthinkable, its very idea dispelled by visible wealth and power, along with palpable powerlessness" (Fredric Jameson, "Marx's Purloined Letter" (1995) in *Valences of the Dialectic* (London: Verso, 2009), 177).

Chapter 5

1. The first part of this phrase is also the title of a book celebrating the company's aesthetic: *Designed by Apple in California* (Cupertino, CA: Apple, Inc., 2016). The book itself is, naturally, printed in China.

2. See, e.g., the discussion of smartphones and the Internet of Things in Adam Greenfield, *Radical Technologies: The Design of Everyday Life* (London: Verso, 2017), 9–62.

3. For an examination of the convergence between the cultural attitudes of the 1960s and the tech industry, see Fred Turner, *From Counterculture to Cyberculture: Stewart Brand, the Whole Earth Network, and the Rise of Digital Utopianism* (Chicago and London: University of Chicago Press, 2006). Interestingly, the last regularly published issue of *Processed World* includes a reasonably favorable review of *WIRED* (alongside a less glowing one of *Mondo 2000*), arguing that "*Wired* is a completely commercial magazine with no political pretensions (other than perhaps that old civil libertarian line), but it manages to provide a lot of useful news and info anyway" (Chris Carlsson, "Tech Talk: Mediamatic and Wired," *Processed World* 32 (Winter/Spring 1993/1994), 65). This assessment seemingly overlooks the more classically libertarian bent of articles like the debut issue's indictment of state schooling (Lewis J. Perelman, "School's Out: Public Education Obstructs the Future, Would You Send Your Kid to a Soviet Collective?," *WIRED* 1.1 (March/April 1993): 71–73 and 104).

4. John Perry Barlow, "A Declaration of the Independence of Cyberspace" (February 8, 1996), archived at https://www.eff.org/cyberspace-independence, accessed July 9, 2021.

5. See Jesse James Garrett, "Ajax: A New Approach to Web Applications," *Adaptive Path*, February 18, 2005, archived at https://web.archive.org/web/20150910072359/http://adaptivepath.org/ideas/ajax-new-approach-web-applications/, accessed July 9, 2021.

6. Ron Eglash, "Computing Power," in *Software Studies: A Lexicon*, ed. Matthew Fuller (Cambridge, MA: MIT Press, 2008), 60.

7. Garrett, "Ajax."

8. With reference to a foundational history of the Internet, Katie Hafner and Matthew Lyon, *Where Wizards Stay Up Late: The Origins of the Internet* (New York: Simon & Schuster, 1996).

9. The inaugural Web 2.0 conference was held in October 2004, although the term had been introduced about five years earlier (see Tim O'Reilly, "What Is Web 2.0: Design Patterns and Business Models for the Next Generation of Software," *O'Reilly*, September 30, 2005, archived at https://www.oreilly.com/pub/a/web2/archive/what-is-web-20.html, accessed July 10, 2021, and Darcy DiNucci, "Fragmented Future,"

Print LIII:IV (July–August 1999): 32 and 221–222). For related, more temporally adjacent, overviews of this period see Wendy Hui Kyong Chun, *Programmed Visions: Software and Memory* (Cambridge, MA: MIT Press, 2010), xi–xii and "Introduction: Did Somebody Say New Media?" in *New Media, Old Media: A Theory and History Reader*, ed. Wendy Hui Kyong Chun and Thomas Keenan (New York: Routledge, 2006), 1–2. Intriguingly from the perspective of 100% utilization, the latter text contends that "new media" as buzzword was largely over by 2003–2004 (2; see also Chun, *Programmed Visions*, 183n1), while the former's diagnosis of the post-dot-com reconstruction's "dim yet gnawing sense that this too will pass" (*xii*) has yet to be borne out. As of this writing the revanchist early twenty-first-century tech status quo has lasted at least four times longer than the comparatively deterritorializing 1990s internet mania.

10. As Anne Pasek argues, "One way to see the rise of cloud computing in recent decades is as a response to the end of Moore's law" (Anne Pasek, "From Atoms to Electrons: An Energy History and Future of Computing," in *Digital Energetics*, ed. Anne Pasek et al. (Lüneberg, Germany; Minneapolis: meson/University of Minnesota Press, 2023), 29). A more nuanced view, however, balances this with additional factors: the popularity of Web 2.0 and later mobile devices due to increased convenience, their corresponding disempowerment of users as a corporate tactic, and the general necessity of concentrating high-intensity computational and network resources in data centers for efficiency (on this last point see also Eglash, "Computing Power," 59–60).

11. See, e.g. Joe Fitzgerald Rodriguez, "Apple, Google Buses Rerouted on Peninsula After Apparent Shooting Damages Windows," *San Francisco Examiner*, January 17, 2018, https://www.sfexaminer.com/news/apple-google-buses-rerouted-on-peninsula-after-apparent-shooting-damages-windows/, accessed July 10, 2021.

12. See, e.g., Joe Vazquez, "Woman Wearing Google Glass Says She Was Attacked in San Francisco Bar," *KPIX 5/CBS SF Bay Area*, February 25, 2014, https://sanfrancisco.cbslocal.com/2014/02/25/woman-wearing-google-glass-says-she-was-attacked-in-san-francisco-bar/, accessed July 10, 2021. As Adam Greenfield notes, the backlash to Glass was almost immediate after its introduction in 2013 (Greenfield, *Radical Technologies*, 84). Google subsequently repositioned the technology as a labor-enhancing tool for custom business applications rather than a potential consumer product.

13. Books in this vein include Jenny Odell, *How to Do Nothing: Resisting the Attention Economy* (Brooklyn, NY: Melville House, 2019) and Wendy Liu, *Abolish Silicon Valley* (London: Repeater, 2020). See also note 65 in chapter 1.

14. Heather Long, "Rent Prices Are Soaring as Americans Flock Back to Cities," *Washington Post*, July 10, 2021, https://www.washingtonpost.com/business/2021/07/09/rent-prices-rising/, accessed July 10, 2021. In a sentiment long familiar to, e.g., Oregonians, one Idaho resident makes his views on the cause of rising housing costs

explicit: "The problem is all these people from California moving in here" (Michael Kobold quoted in Long, "Rent Prices Are Soaring").

15. Mark Gurman, "Apple Decentralizes from Silicon Valley, but Workers Just Want to Be Remote," *Bloomberg*, July 4, 2021, https://www.bloomberg.com/news/newsletters/2021-07-04/apple-hq-is-no-longer-the-center-of-the-company-s-universe-kqpi2879, accessed July 10, 2021.

16. See, e.g., Andrew Lison, "Hardware Standardization and State-Socialist Piracy: The Global Reach of the Zilog Z80," *IEEE Annals of the History of Computing* 46, no. 2 (April–June 2024): 40–41, 49.

17. Christophe Lécuyer, *Making Silicon Valley: Innovation and the Growth of High-Tech, 1930–1970* (Cambridge, MA: MIT Press, 2006), 40, 41.

18. Ibid., 82.

19. Ibid., 164 (see also the previous page and 299–300).

20. Ibid., 164, 299.

21. Ibid., 341n94.

22. Ibid., 204.

23. Christophe Lécuyer and David C. Brock, *Makers of the Microchip: A Documentary History of Fairchild Semiconductor* (Cambridge, MA: MIT Press, 2010), 120 (see the previous page for a reproduction of company cofounder Jay Last's notes on the topic).

24. Lécuyer, *Making Silicon Valley*, 204, 207. See also Brock and Lécuyer, *Makers of the Microchip*, 234, Aaron Benanav, *Automation and the Future of Work* (London: Verso, 2019), 27, and Lisa Nakamura, "Indigenous Circuits: Navajo Women and the Racialization of Early Electronic Manufacture," *American Quarterly* 66, no. 4 (December 2014): 922.

25. Ibid., 919–941 (quote from 923).

26. Ibid., 923. Lécuyer argues that, thanks to the lower pay its expansions from 1962 onward made possible, "Fairchild Semiconductor became less of a military contractor and more of a civilian manufacturer" (Lécuyer, *Making Silicon Valley*, 207; see also Nakamura, "Indigenous Circuits," 922).

27. Ibid., 935.

28. Ibid., 926. We might note here, however, that this concept is not unique to twentieth- and twenty-first-century labor but can arguably be found in Marx: "The flexibility of capital, its indifference to the particular forms of the labour process it acquires, is extended by capital to the worker. He is required to be capable of the same flexibility or *versatility* in the way he applies his labour-power . . . the capitalist mode of production itself raises obstacles in the way of its own tendency,

but it pushes to one side all *legal* and other *extra-economic* obstructions standing in the way of this versatility" (Karl Marx, "Results of the Immediate Process of Production" (1863–1866) in *Capital: A Critique of Political Economy*, vol. 1 (1867) trans. Ben Fowkes (1976) (London: Penguin, 1990), 1013 (emphases in original)). That the German word translated here as "flexibility" is *Flüssigkeit*, or "fluidity," rather than the more literal *Biegsamkeit* should not detract from the larger point that capital has long been happy to bend superstructural concerns to its material needs.

29. Sean Cubitt, *Finite Media: Environmental Implications of Digital Technologies* (Durham, NC: Duke University Press, 2017), 14. For how this unfolds in other contexts see, e.g., 94–95 and Greenfield, *Radical Technologies*, 19.

30. Nakamura, "Indigenous Circuits," 932.

31. Ibid., 924 (emphasis in original).

32. Helen Hester, *Xenofeminism* (Cambridge, UK: Polity, 2018), 8.

33. Nakamura, "Indigenous Circuits," 937–938.

34. Ibid., 938.

35. Karl Marx, *Grundrisse: Foundations of the Critique of Political Economy (Rough Draft)* (1857–1858) trans. Martin Nicolaus (1973) (London: Penguin, 1993), 692.

36. Stephanie Yang, "Chip Makers Contend for Talent as Industry Faces Labor Shortage," *Wall Street Journal*, January 2, 2022, available online at https://www.wsj.com/tech/chip-makers-contend-for-talent-as-industry-faces-labor-shortage-11641124802, accessed August 21, 2025.

37. See note 17 in chapter 1.

38. Laboria Cuboniks, *The Xenofeminist Manifesto: A Politics for Alienation* (2015), (London: Verso, 2018), 19 (§0x03; emphasis mine).

39. Lisa Nakamura, "Indigenous Circuits," *Computer History Museum Blog*, January 2, 2014 https://computerhistory.org/blog/indigenous-circuits/, accessed November 5, 2022.

40. Ibid.

41. Probir K. Bondyopadhyay, "W=Shockley, the Transistor Pioneer—Portrait of an Inventive Genius," *Proceedings of the IEEE* 86, no. 1 (January 1998): 196. A reassessment of Shockley within the scientific community is ongoing, e.g. H. Holden Thorp, "Shockley Was a Racist and Eugenicist," *Science* vol. 378, no. 6621 (November 17, 2022): 683, available online at https://www.science.org/doi/10.1126/science.adf8117, accessed August 21, 2025).

42. David C. Brock, "From Automation to Silicon Valley: The Automation Movement of the 1950s, Arnold Beckman, and William Shockley," *History and Technology*

28, no. 4 (December 2012): 375–401. On Beckman's background see e.g., 385–389; on cybernetics, 376–377. See also Arnold Thackray, "Before Moore's Law: Lineages of Chemistry and Electricity," in *Understanding Moore's Law: Four Decades of Innovation*, ed. David C. Brock (Philadelphia: Chemical Heritage Press, 2006), 5–6, and David C. Brock, "The Backdrop to Moore's Law: Developments in Semiconductor Electronics To 1965," in *Understanding Moore's Law*, ed. Brock, 13–14.

43. Brock, "From Automation to Silicon Valley," 383–384.

44. Ibid., e.g., 376–378, 396–397. On Wiener and/or the automatic factory, see e.g., 376, 378, 397, and note 93 in chapter 2.

45. Quoted in Brock, "From Automation to Silicon Valley," 387. This is not to suggest that Beckman was a left utopian by any means. He would go on, for example, to anticipate the *WIRED* line on education (see note 3 earlier), ironically blaming "the broad socialistic movement" in the United States for "the current wastage of potential technical manpower" compared to the Soviet Union (Dr. Arnold O. Beckman, "A Businessman's View on the 'Failure' of Education," *U.S. News & World Report*, November 30, 1956: 83–86, 89 (quotes from 86, 89)).

46. William Shockley, "A.T.R. Project" (1952), William Shockley Papers (SC0222), Dept. of Special Collections and University Archives, Stanford University Libraries, available online at https://searchworks.stanford.edu/view/pd057vc3349, accessed August 16, 2024 (also quoted and discussed in Brock, "From Automation to Silicon Valley," 382–383).

47. Brock, "From Automation to Silicon Valley," 383. See also William Shockley, "A.T.R. Project" (1952), William Shockley Papers (SC0222), Dept. of Special Collections and University Archives, Stanford University Libraries, available online at https://searchworks.stanford.edu/view/cm000qw0881, accessed August 16, 2024.

48. Brock, "From Automation to Silicon Valley," 380–383. His initial plan "to form a new start-up company . . . and then perhaps to approach a large established company about acquiring" it, Brock notes, "was a strategy that would become commonplace four decades later in Silicon Valley's dotcom boom" (382; see also 391).

49. Ibid., 383.

50. Ibid., 390 (emphasis in original), 396.

51. Ibid., 396. On automating transistor production see also 391–392 and note 47 in chapter 2. Ironically, what would enable the anti-socialist Beckman and Shockley to go into business together, and "the very possibility for a semiconductor industry," was state intervention, in the form of a trust-busting agreement opening some phone company intellectual property, including transistor patents, to reasonable licensing terms (Brock, "From Automation to Silicon Valley," 392).

52. J. Bardeen and W. H. Brattain, "The Transistor, A Semi-Conductor Triode," *Physical Review* 74, no. 2 (July 1948): 230–231. For the elemental composition of the transistor, see 230 and the source they point to, Henry C. Torrey and Charles A. Whitmer, *Crystal Rectifiers* (New York and London: McGraw-Hill, 1948), especially 365. For an accessible overview of the history of semiconductor engineering see Brock, "The Backdrop to Moore's Law," 7–21.

53. W. Shockley, "The Theory of *p-n* Junctions in Semiconductors and *p-n* Junction Transistors," *The Bell System Technical Journal* 28, no. 3 (July 1949): 435–489. See also Brock, "The Backdrop to Moore's Law," 11, Bondyopadhyay, "W=Shockley," 191–192.

54. W. Shockley, "Possible Transfer of Metallurgical and Astronomical Approaches to Problem of Environment versus Ethnic Heredity," *Science* 154, no. 3747 (October 21, 1966): 428.

55. E.g., Shun-ichi Maekawa, "Diffusion of Phosphorus into Silicon," *J. of the Physical Society of Japan* 17, no. 10 (1962): 1592–1597, Bondyopadhyay, "W=Shockley," 199.

56. Shockley, "Possible Transfer of Metallurgical and Astronomical Approaches," 428. As his title suggests, the Cold War was another factor Shockley used to sell his approach: "Data about such individuals, whose prior development is uninfluenced by the study, are analogous to data of pre-sputnik astronomy" (ibid.).

57. William Shockley, "On the Statistics of Individual Variations of Productivity in Research Laboratories," *Proceedings of the IRE* 45, no. 3 (March 1957): 279–290.

58. Ibid., 285.

59. Ibid., 286.

60. Ibid.

61. Shockley, "Possible Transfer of Metallurgical and Astronomical Approaches," 428.

62. S. L. Washburn, "The Study of Race," *American Anthropologist* New Series, 65, no. 3, Part 1 (June 1963): 521–531. On IQ tests see, e.g., Ajitha Reddy, "The Eugenic Origins of IQ Testing: Implications for Post-Atkins Litigation," *DePaul Law Review* 57, no. 3 (Spring 2008): 667–677.

63. Shockley, "Possible Transfer of Metallurgical and Astronomical Approaches," 428.

64. William Shockley quoted in "Is Quality of U.S. Population Declining? Interview with a Nobel Prize-Winning Scientist," *U.S. News & World Report*, November 22, 1965, 70.

65. Ibid., 71.

66. Ibid.; Beckman, "A Businessman's View."

67. "Is Quality of U.S. Population Declining?," 70, 71.

68. See, e.g., Thorp, "Shockley Was A Racist and Eugenicist."

69. "William Shockley," *Southern Poverty Law Center* website https://www.splcenter.org/resources/extremist-files/william-shockley/; Associated Press, "Shockley, Nobel Winner, Files for Senate Race in California," published in *The New York Times*, February 12, 1982, A16, available online at https://www.nytimes.com/1982/02/12/us/shockley-nobel-winner-files-for-senate-race-in-california.html; Wolfgang Saxon, "Obituary: William B. Shockley, 79, Creator of Transistor and Theory on Race," *New York Times*, August 14, 1989, available online at https://archive.nytimes.com/www.nytimes.com/learning/general/onthisday/bday/0213.html (all URLs accessed August 21, 2025).

70. See Laura Feiveson, "Labor Unions and the U.S. Economy," *U.S. Department of the Treasury* August 28, 2023, https://home.treasury.gov/news/featured-stories/labor-unions-and-the-us-economy. accessed January 29, 2025: "Union membership peaked in the 1950s at one-third of the workforce. At that time, despite pervasive racial and gender discrimination, overall income inequality was close to its lowest level since its peak before the Great Depression, and was continuing to fall."

71. As Reddy notes, citing Nicholas Lemann, *The Big Test: The Secret History of the American Meritocracy* (New York: Farrar, Straus and Giroux, 2000), 344, "By the 1950s, intelligence testing became a routine facet of work and school life nationwide" (Reddy, "The Eugenic Origins of IQ Testing," 673). See 673–674 for a critique of American meritocracy.

72. Also lampooned by the television show *Silicon Valley*, in its portrayal of a founder's psychedelic mushroom trip/"vision quest" (another appropriation of Native American culture!) spent repeating the slogan "Making the World a Better Place" over and over again. See also Ruha Benjamin, *Race After Technology: Abolitionist Tools for the New Jim Code* (Cambridge, UK: Polity, 2019), 150–151, 213n36.

73. One of the unanswered questions from Nakamura's initial inquiry is "whether [Shockley's] opinions influenced his hiring decisions, and by extension, those of other CEO's in Silicon Valley's formative years. Could it be possible that the Asian 'model minority' stereotype, and the scarcity of African American and Latinos in the computer industry, might have something to do with hiring practices based on opinions like Shockley's?" (Nakamura, "Indigenous Circuits" blog post). See also Benjamin, *Race After Technology*, 58

74. Michel Foucault, *The Birth of Biopolitics: Lectures at the Collège de France, 1978–1979*, ed. Michel Senellart (2004), trans. Graham Burchell (Basingstoke, UK: Palgrave Macmillan, 2008), 317; Michel Foucault, *"Society Must Be Defended": Lectures at the Collège de France, 1975–1976*, ed. Mauro Bertani and Alessandro Fontana (1997), trans. David Macey (New York: Picador, 2003), 246.

75. Ibid., 255. He elaborates, "When I say 'killing,' I obviously do not mean simply murder as such, but also every form of indirect murder: the fact of exposing someone to death, increasing the risk of death for some people, or, quite simply, political death, expulsion, rejection, and so on" (256). See also the discussion of eugenics as Foucauldian biopower/biopolitics and genetics as a precursor to the *logos* of computer programs in Chun, *Programmed Visions*, 119–129.

76. Craig Wilse, "Surplus Life: Biopower and Neoliberalism," *Scholar & Feminist Online* 11, no. 1–2 (Fall 2012/Spring 2013), https://sfonline.barnard.edu/surplus-life-biopower-and-neoliberalism/, accessed August 21, 2025.

77. Foucault, *The Birth of Biopolitics*, 203–205. This is a further twist to the abandonment of causality we have first encountered in quantum physicists. Aaron Benanav (discussed later in this chapter) makes the slightly different argument that "starting in the 1970s . . . unemployment insurance programs went into crisis: they had been designed for short bouts of cyclical unemployment in fast-growing economies, not long-term unemployment in stagnating economies. Active labor market policies replaced passive income-support systems as the main response to job loss" (Aaron Benanav, *Automation and the Future of Work* (London: Verso, 2019), 46–47). Regardless, Benanav and Foucault agree that emphasizing the short-term qualities of unemployment benefits was the de facto neoliberal model.

78. Foucault, *"Society Must Be Defended,"* 258.

79. Elisa (only subjects' first names are given) quoted in Mary Beth Meehan and Fred Turner, *Seeing Silicon Valley: Life inside a Fraying America* (2018) (Chicago: University of Chicago Press, 2021), 96 (emphasis in original). Meehan reports that Elisa was ultimately assuaged by Zuckerberg's response "to invest in the local schools and to make other contributions to the area" (ibid.).

80. Wilse, "Surplus Life." See also Greenfield, *Radical Technologies*, 190.

81. Benjamin, *Race After Technology*, 115.

82. Chun, *Programmed Visions*, 124.

83. Brock, "From Automation to Silicon Valley," 378.

84. Bill Joy, "Why the Future Doesn't Need Us," *WIRED* 8.04 (April 2000), available online at https://www.wired.com/2000/04/joy-2/, accessed January 20, 2022.

85. Ibid.

86. Kevin Kelly quoted on The Singularity Summit at Stanford website, 2006, archived at https://web.archive.org/web/20060423160955/http://sss.stanford.edu/, accessed January 18, 2022 (emphases in original).

87. Ray Kurzweil, *The Singularity Is Near: When Humans Transcend Biology* (New York: Viking, 2005), 67.

88. Ibid. Since electromechanical computers typically used relay switches the first category is perhaps more properly understood, as the icon Kurzweil includes of a punch card suggests, as simply mechanical.

89. Ray Kurzweil, *The Age of Intelligent Machines* (Cambridge, MA: MIT Press, 1990), 419. As we have seen in chapter 4, however, the most significant developments in AI ultimately came about precisely through the more-specialized GPU.

90. Ray Kurzweil, "Ask Ray | The Future of Moore's Law," (May 28, 2013), https://www.thekurzweillibrary.com/ask-ray-the-future-of-moores-law, accessed August 21, 2025. For more on the barriers to molecular computing, see Kevin F. Kelly (not the *WIRED* maverick) and Cyrus C. M. Mody, "Whatever Happened to the Molecular Computer?," *IEEE Spectrum*, September 25, 2015, https://spectrum.ieee.org/whatever-happened-to-the-molecular-computer, accessed January 31, 2022. They note, "The story of molecular electronics is special for what it says about the nature of research and progress, and about the human capacity for optimism and hubris" (ibid.).

91. Kurzweil, *The Singularity Is Near*, 66.

92. The key text here is Kodwo Eshun, *More Brilliant than the Sun: Adventures in Sonic Fiction* (London: Quartet, 1998), but for more on this context see Andrew Lison, "New Media, 1989: Cubase and the New Temporal Order," *Computational Culture* 8 (July 2021), http://computationalculture.net/new-media-1989-cubase-and-the-new-temporal-order/ and Benjamin Noys, "Dance and Die: Obsolescence and Embedded Aesthetics of Acceleration," in *Hopeless Youth!* ed. Francisco Martínez and Pille Runnel (Tartu: Estonian National Museum, 2015), 293–296.

93. Robin Mackay and Armen Avanessian, "Introduction" in *#ACCELERATE#: The Accelerationist Reader*, ed. Robin Mackay and Armen Avanessian (London: Urbanomic, 2014), 4.

94. Alex Williams and Nick Srnicek, "#Accelerate: Manifesto for an Accelerationist Politics" (2013), in *#ACCELERATE#*, ed. Mackay and Avanessian, 356.

95. Aaron Bastani, *Fully Automated Luxury Communism: A Manifesto* (London: Verso, 2019), 45.

96. Karl Marx, "Preface to A Contribution to the Critique of Political Economy" (1859), in *Early Writings*, trans. Rodney Livingstone and Gregor Benton (London: Penguin, 1992), 426, 425.

97. E.g., Max Horkheimer and Theodor W. Adorno, *Dialectic of Enlightenment: Philosophical Fragments* (1944/1947), trans. Edmund Jephcott, ed. Gunzelin Schmid Noerr (Stanford, CA: Stanford University Press, 2002).

98. See, e.g., Mark Fisher, *Capitalist Realism: Is There No Alternative?* (Winchester, UK: Zero Books, 2009) and fellow-traveler Simon Reynolds's *Retromania: Pop Culture's Addiction to Its Own Past* (New York: Faber and Faber, 2011). Critics have likewise pointed out that twenty-first-century accelerationism does not have recourse to the kind of era-defining musical innovation that the CCRU itself did (Noys, "Dance and Die," 301–302).

99. See Karl Marx, *Capital: A Critique of Political Economy*, vol. 3 (1894), trans. David Fernbach (1981) (London: Penguin, 1991), 315–375.

100. Benanav, *Automation and the Future of Work*, 32–33.

101. Ibid., 3. These internecine debates on the left build off a number of more mainstream considerations of automation that emerged in the early 2010s, especially Erik Brynjolfsson and Andrew McAfee, *The Second Machine Age: Work, Progress, and Prosperity in a Time of Brilliant Progress* (New York: W.W. Norton & Co., 2014).

102. Benanav, *Automation and the Future of Work*, 40, 116n34–35.

103. Ibid., 18. See also note 77 in chapter 2.

104. Benanav, *Automation and the Future of Work*, 86 (see also 12, 89–90).

105. Karl Marx, *Capital: A Critique of Political Economy*, vol. 1 (1867), trans. Ben Fowkes (1976) (London: Penguin, 1990), 515n33. See also Benanav, *Automation and the Future of Work*, 41.

106. Ibid., 34.

107. Marx, *Capital* vol. 3, 318.

108. Ibid., 319.

109. Ibid., 322. See also the following page.

110. Ibid., 324 (emphases in original).

111. Ibid., 325.

112. Ibid., 329. This is a more limited—but real—version of the decommodifying impact that advocates of 3D printers claim they could one day potentially have (Greenfield, *Radical Technologies*, 90).

113. Marx, *Capital* vol. 3, 360.

114. Ibid., 360–362.

115. Ibid., 362–363.

116. Ibid., 362.

117. Ibid., 315, 375. See also Ernest Mandel, "Introduction" to Marx, *Capital* vol. 3, 51.

118. Paolo Virno, "Notes on the 'General Intellect'" (1990/1992), in *Marxism Beyond Marxism* trans. Cesare Casarino, ed. Saree Makdisi, Cesare Casarino, and Rebecca E. Karl (New York: Routledge, 1996), 266. These approaches' interpretation of Marx's "Fragment on Machines" can often be traced back to the reading in Antonio Negri, *Marx Beyond Marx (Lessons on the Grundrisse)* (1978/1979), trans. Harry Cleaver, Michael Ryan, and Maurizio Viano, ed. Jim Fleming (1984) (Brooklyn: Autonomedia, 1991), 137–150. See also Mark Alizart, *Cryptocommunism*, trans. Robin Mackay (Cambridge, UK: Polity, 2020), 84–85 and note 25 in chapter 6.

119. Marx, *Capital* vol. 3, 360.

120. Ibid., 362.

121. Ibid., 364.

122. On this latter means of accumulation, Marx writes that when the profit rate declines, "we have swindling and general promotion of swindling, through desperate attempts in the way of new methods of production, new capital investments and new adventures, to secure some kind of extra profit" (ibid., 367).

123. Benanav, *Automation and the Future of Work*, 28. He would thus seem to agree with Brynjolfsson's redistributive thesis discussed in chapter 2: "In an environment where average growth rates are low, firms can only achieve high rates of growth by taking market share from their competitors" (ibid.; see also 26).

124. Marx, *Capital* vol. 3, 363–364.

125. Ibid., 339. Jonathan Beller elaborates: "As labor power makes up an increasing smaller proportion of the total value of capital outlay . . . the return on the cost of labor power must increase. . . . If not, the rate of profit falls. Thus, the drive to innovation is a drive to increase the efficiency of labor" (Jonathan Beller, *The World Computer: Derivative Conditions of Racial Capitalism* (Durham, NC: Duke University Press, 2021), 30; see also note 23 in chapter 6).

126. See note 46 in chapter 1.

127. Marx, *Capital* vol. 3, 339. See also note 56 in chapter 2.

128. See US Department of Commerce, "Results from Semiconductor Supply Chain Request for Information" blog post, January 25, 2022, https://www.commerce.gov/news/blog/2022/01/results-semiconductor-supply-chain-request-information, accessed January 26, 2022. According to this report, the integrated circuit industry approached its own, commodity production, version of 100% utilization during this time: amongst "major supply and demand mismatch . . . from Q2 of 2020 through 2021, semiconductor fabs operated at over 90% utilization, which is incredibly high for a production process that requires regular maintenance and very high amounts of energy" (ibid.).

129. See, e.g., Benanav, *Automation and the Future of Work*, *x*, 34, 56–60, 49–50.

130. Herb Sutter, "The Free Lunch Is Over: A Fundamental Turn Toward Concurrency in Software" (December 2004/2009) http://www.gotw.ca/publications/concurrency-ddj.htm accessed April 3, 2017, also published in *Dr. Dobb's Journal* 30, no. 3 (March 2005): 16–22.

131. Moritz Lipp et al., "Meltdown: Reading Kernel Memory from User Space," *27th USENIX Security Symposium* Conference Proceedings (August 15–17, 2018), 973–990 and Paul Kocher et al., "Spectre Attacks: Exploiting Speculative Execution," *2019 IEEE Symposium on Security and Privacy* Conference Proceedings (May 19–23, 2019), 1–19. While the exact specifications and presentation dates vary, with Meltdown focusing on out-of-order execution of explicitly specified instructions and Spectre targeting speculative execution, or the carrying out of *potentially* desirable instructions, both vulnerabilities were announced simultaneously, by many of the same researchers, in early 2018. Both papers make clear the features they exploit were implemented for performance reasons.

132. Matt Spencer, "Creative Malfunction: Finding Fault with Rowhammer," *Computational Culture* 8 (July 2021), http://computationalculture.net/creative-malfunction-finding-fault-with-rowhammer/, accessed February 5, 2022. Variations on this exploit continue to be found (e.g., Dan Goodin, "There's a New Way to Flip Bits in DRAM, and It Works Against the Latest Defenses," *Ars Technica*, October 19, 2023, https://arstechnica.com/security/2023/10/theres-a-new-way-to-flip-bits-in-dram-and-it-works-against-the-latest-defenses/, accessed August 15, 2024 and Nureddin Kamadan et al., "ECC.fail: Mounting Rowhammer Attacks on DDR4 Servers with ECC Memory," *Proceedings of the 34th USENIX Security Symposium* (August 13–15, 2025), available online at https://www.usenix.org/system/files/usenixsecurity25-kamadan.pdf, accessed August 21, 2025).

133. See, e.g., Joel Hruska, "Intel Performance Hit 5x Harder Than AMD After Spectre, Meltdown Patches," *ExtremeTech*, May 20, 2019, https://www.extremetech.com/internet/291649-intel-performance-amd-spectre-meltdown-mds-patches, accessed June 13, 2023, and Dan Goodin, "Ubuntu Disables Intel GPU Security Mitigations, Promises 20% Performance Boost," *Ars Technica*, June 25, 2025, https://arstechnica.com/security/2025/06/ubuntu-disables-intel-gpu-security-mitigations-promises-20-performance-boost/, accessed August 21, 2025.

134. Ken Thompson, "Reflections on Trusting Trust," *Communications of the ACM* 27, no. 8 (August 1984): 761–763. For a detailed examination, see Ben Allen, "Source Code Isn't" in *Your Computer Is on Fire,* eds. Thomas S. Mullaney et al. (Cambridge, MA: MIT Press, 2021), 273–295. As he argues, "The transition between low-level and high-level languages is . . . the first moment that software developers can begin to treat the machines that subtend software as irrelevant" (278).

135. Contrast this with the early-2000s characterization that "while a burglar's exploits often rely on physical force . . . a hacker's exploits generally rely on *logical* force" (Alexander R. Galloway, *Protocol: How Control Exists After Decentralization* (Cambridge, MA: MIT Press, 2004), 168 (emphasis in original)).

136. Fredric Jameson, "Three Names of the Dialectic," in *Valences of the Dialectic* (London: Verso, 2009), 46.

137. Benanav, *Automation and the Future of Work*, 82.

138. Karl Marx, *Capital: A Critique of Political Economy*, vol. 1 (1867), trans. Ben Fowkes (1976) (London: Penguin, 1990), 133.

139. E.g., Noys, "Dance and Die," 298, 294–296. I discuss Noys further in the following chapter.

140. Liner notes to Ricardo Villalobos, *Alcachofa* (Playhouse PLAY083, 2003).

Chapter 6

1. *Oxford English Dictionary*, s.v. "robot $(n)^1$" and "robot $(n)^2$."

2. Benjamin Noys, "Dance and Die: Obsolescence and Embedded Aesthetics of Acceleration," in *Hopeless Youth!* ed., Francisco Martínez and Pille Runnel (Tartu: Estonian National Museum, 2015), 297.

3. Ibid., 306. For an in-depth exploration of this issue in terms of uneven development and the impact of globalization, particularly outside of the West, see Anna Lowenhaupt Tsing, *Friction: An Ethnography of Global Connection* (Princeton: Princeton University Press, 2004). See also note 56 later.

4. Norbert Wiener, *Cybernetics; or, Control and Communication in the Animal and Machine* (1948), 2nd ed. (1961) (Cambridge, MA: MIT Press, 2019).

5. Ibid., 156.

6. Cybernetics famously takes its name from the Greek word for "steersman," from which the English word "governor" is also descended. The latter term has a rich history in thermodynamics, to which cybernetics is itself deeply indebted, as Wiener makes clear (ibid., 18). See also note 102 in chapter 4.

7. The intellectual lineage running from Maxwell's demon, as a thought experiment countering the second law of thermodynamics with which Wiener was obsessed, to background running UNIX processes known as daemons, is one particularly significant example of these cybernetic interchanges, examined as such in Fenwick McKelvey, *Internet Daemons: Digital Communications Possessed* (Minneapolis: University of Minnesota Press, 2018), e.g., 23–43.

8. Fred Turner, *From Counterculture to Cyberculture: Stewart Brand, the Whole Earth Network, and the Rise of Digital Utopianism* (Chicago and London: University of Chicago Press, 2006), 21.

9. Philip Mirowski, *Machine Dreams: Economics Becomes a Cyborg Science* (Cambridge, UK: Cambridge University Press, 2001), 43.

10. Anson Rabinbach, *The Human Motor: Energy, Fatigue, and the Origins of Modernity* (New York: Basic Books, 1990), 47–48. See also 63.

11. Ibid., 4. As he notes, analogies between the body and a machine on the one hand, and a factory, as Wiener draws, on the other, have classical and sixteenth-century roots, respectively (51). See also the discussion of Rabinbach, work, and the second law of thermodynamics in Mary Ann Doane, *The Emergence of Cinematic Time: Modernity, Contingency, the Archive* (Cambridge, MA: Harvard University Press, 2002), 114–117.

12. Rabinbach, *The Human Motor*, 55, 59 (citing H. Helmholtz, "Über die Erhaltung der Kraft" (1862–1863), in H. Helmholtz, *Populäre wissenschaftliche Vorträge*, vol. 2 (Braunschweig: Friedrich Vieweg und Sohn, 1871), 137–179). As Rabinbach notes (Rabinbach, *The Human Motor*, 313n38), Thomas Kuhn connects Helmholtz's use of *Kraft*, or power, with a German translation of Sadi Carnot's *puissance motrice* ("motive power"), though Kuhn also argues that "the parallels in the French and English engineering traditions" for terms like *Arbeitskraft* "are obvious" (Thomas Kuhn, *The Essential Tension: Selected Studies in Scientific Tradition and Change* (Chicago: University of Chicago Press, 1977), 88n48). See also my discussion of Carnot in chapter 3.

13. Rabinbach, *The Human Motor*, 61. See also note 99 in chapter 3.

14. Rabinbach, *The Human Motor*, 79. For him, "Marx's phenomenology of labor power is not only contemporary with the Helmholtzian revolution in scientific perception; it is a direct consequence of it" (81). For a more recent accelerationist consideration of Marxism as "an extension of the laws of thermodynamics to society and the economy" see Mark Alizart, *Cryptocommunism*, trans. Robin Mackay (Cambridge, UK: Polity, 2020), e.g., 48–58 (quote from 49). For a critique, see note 25 later.

15. Rabinbach, *The Human Motor*, 80. See also Alizart, *Cryptocommunism*, 51–54 and notes 23 and 25 later.

16. Rabinbach, *The Human Motor*, 73. It is important to note that, for Marx, it is not only labor but the natural world that forms a "source of material wealth, i.e. of the use values it produces" (Karl Marx, *Capital: A Critique of Political Economy*, vol. 1 (1867), trans. Ben Fowkes (1976) (London: Penguin, 1990), 134; see also Karl Marx, "Critique of the Gotha Programme (1875)" trans. Joris de Bres (1974) in Karl Marx, *The First*

International and After (*Political Writings*, vol. 3) ed. David Fernbach (1974) (London: Verso, 2010), 341). However, this is obviously not the same as saying that a machinic process creates value.

17. Rabinbach, *The Human Motor*, 4.

18. Marx, *Capital* vol. 1, 270.

19. Norbert Wiener, *The Human Use of Human Beings: Cybernetics and Society* (1950), Revised Edition (1954) (Cambridge, MA: Da Capo Press, 1988), 12.

20. Orit Halpern, *Beautiful Data: A History of Vision and Reason Since 1945* (Durham, NC: Duke University Press, 2014), 40–41.

21. Wiener, *Cybernetics*, 220; Mirowski, *Machine Dreams*, 12 (on this Wiener quote see 64).

22. Wiener, *Cybernetics*, 221.

23. Alizart, *Cryptocommunism*, 82–84, understands the falling rate of profit cycle, the "cost of innovation" (82; see also note 125 in chapter 5) in which "each new cycle comes out of the previous crisis stronger" (84), similarly. McKelvey perceptively connects Wiener's notion of metastability with Gilles Deleuze's use of the term to characterize the post-industrial replacement of strict Foucauldian discipline with a more flexible mode of authority (McKelvey, *Internet Daemons*, 42; see also 28–29 and Gilles Deleuze, "Postscript on the Societies of Control," *October* 59 (Winter 1992): 3–7). It is in this sense that neoliberal control societies seek to govern the disruption produced by industrial economic cycles.

24. See, e.g., Alizart, *Cryptocommunism*, 107. For a description of the adoption of feedback from early cybernetics into economics see Mirowski, *Machine Dreams*, 205–206.

25. Alizart, *Cryptocommunism*, 57. Alizart draws, like the accelerationists outlined previously (see note 118 in chapter 5), on the *Grundrisse*, arguing that "at least according to the theory of 'General Intellect,' Marxism's success depends directly upon the progress of science, technology and, in particular, Moore's law" (89–90). His reading of the falling rate of profit—that "fixed capital acts as an 'entropy sensor,' and this is the secret of the miraculous survival of capitalism" (57; see also note 15 earlier), necessarily raises the question of what happens when this warning system can no longer be improved.

26. Kraftwerk, "The Man–Machine," *The Man–Machine* (Capitol E-ST 11728, 1978). On the group's influence on techno, see, e.g., Simon Reynolds, *Energy Flash: A Journey Through Rave Music and Culture* (1998) updated 20th anniversary edition (London: Picador, 2008), 2–6.

27. Kraftwerk, "Die Mensch–Maschine," *Die Mensch–Maschine* (EMI Electrola/Kling Klang 1C 058–32 843, 1978).

28. See William R. Bennett, "Secret Telephony as a Historical Example of Spread-Spectrum Communication," *IEEE Transactions on Communications* 31, no. 1 (January 1983): e.g. 98–99, *A History of Engineering and Science in the Bell System,* vol. II (*National Service in War and Peace (1925–1975)*), ed. M. D. Fagen (Murray Hill, NJ: Bell Telephone Laboratories, 1978), 296–316, and Andrew Hodges, *Alan Turing: The Enigma* (1983) (Princeton, NJ: Princeton University Press, 2014), 310–311 (which also draws on Fagen, ed., *A History of Engineering and Science in the Bell System* vol. II). For a popular history focused less on the digital than the acoustic and subsequent musical aspects, see Dave Tompkins, *How to Wreck a Nice Beach: The Vocoder from World War II to Hip-Hop—The Machine Speaks* (2010) (Chicago/Brooklyn: Stop Smiling/Melville House, 2011), e.g., 52–73.

29. Bennet, "Secret Telephony," 98. See also Fagen ed., *History of Engineering and Science in the Bell System* vol. II, 296–297, Tompkins, *How to Wreck a Nice Beach*, 19, 59, and 55, and Hodges, *Alan Turing*, 310–311.

30. G. S. Vernam, "Cipher Printing Telegraph Systems for Secret Wire and Radio Telegraphic Communications," *Transactions of the American Institute of Electrical Engineers* XLV (January–December 1926): 295–301. See also Fagen ed., *History of Engineering and Science in the Bell System* vol. II, 244–246, Bennett, "Secret Telephony," 98 and Hodges, *Alan Turing*, 288–289 and 310–311.

31. Friedrich Kittler, "Code (or, How You Can Write Something Differently)" (2003), in *Software Studies: A Lexicon*, trans. Tom Morrison and Florian Cramer, ed. Matthew Fuller (Cambridge, MA: MIT Press, 2008), 41.

32. Vernam, "Cipher Printing Telegraph Systems," 299.

33. See Hodges, *Alan Turing*, 309–316, Fagen ed., *History of Engineering and Science in the Bell System* vol. II, 316–317, Tompkins, *How to Wreck a Nice Beach*, 59–60, and Mirowski, *Machine Dreams*, 68. Kittler's analysis, which takes a few liberties with the details, is also worth noting (Friedrich A. Kittler, *Gramophone, Film, Typewriter* (1986), trans. Geoffrey Winthrop-Young and Michael Wutz (Stanford, CA: Stanford University Press, 1999), 49). On the postmodern aspects of informational culture see Tiziana Terranova, *Network Culture: Politics for the Information Age* (London: Pluto, 2004), 6–38.

34. Kittler, "Code," 41.

35. Claude E. Shannon, "Communication Theory of Secrecy Systems," *The Bell System Technical Journal* 28, no. 4 (October 1949): 656–715. See also Claude E. Shannon, "The Mathematical Theory of Communication" (1948), in Claude E. Shannon and Warren Weaver, *The Mathematical Theory of Communication* (1948/1949) (Urbana, IL: University of Illinois Press, 1998), 29–125, Fagen ed., *History of Engineering and Science in the Bell System* vol. II, 315–317, Hodges, *Alan Turing*, 694nBP8, Mirowski, *Machine Dreams*, 69, and Shawn James Rosenheim, *The Cryptographic*

Imagination: Secret Writing from Edgar Poe to the Internet (Baltimore: Johns Hopkins University Press, 1997), 192–193.

36. Shannon, "Mathematical Theory of Communication," 49. For the proof see Shannon, "Communication Theory of Secrecy Systems," 679–683. See also Fagen ed., *History of Engineering and Science in the Bell System* vol. II, 301.

37. Hodges, *Alan Turing*, 289 (see also the previous page).

38. Bennet, "Secret Telephony," 98; Tompkins, *How to Wreck a Nice Beach*, 70.

39. Jacques Derrida, "Signature Event Context (1972)," trans. Samuel Weber and Jeffrey Mehlman (1977), in Jacques Derrida, *Limited Inc.* (Evanston: Northwestern University Press, 1988), 17.

40. Wiener, *Cybernetics*, 94.

41. Bennett, "Secret Telephony," 98–99. See also Fagen ed., *History of Engineering and Science in the Bell System* vol. II, 302–305 and Hodges, *Alan Turing*, 310–311.

42. Wiener, *Cybernetics*, 85, 86.

43. Tompkins, *How to Wreck a Nice Beach*, 187.

44. See Tilman Baumgärtel, *Now and Forever: Towards a Theory and History of the Loop* (Winchester, UK: Zero Books, 2023), 307–325 for a reading of techno as homeostatic in the positive sense. Where he argues that, alongside minimal composition and other forms of "electronic dance music . . . at a symbolic level, this kind of music is . . . about what a more constructively informed relationship between humans and machines might look like" (310), my argument here is, firstly, that techno is more ambivalent; and secondly and more importantly, that this relationship will ultimately have to be realized, at the level of the material, beyond any cybernetic relation.

45. Tompkins, *How to Wreck a Nice Beach*, 176.

46. Cybotron, *Enter* (Fantasy F-9625, 1983).

47. Cybotron, *Interface: The Roots of Techno* (Southbound CDSEWD 069, 1994).

48. Ibid.

49. Tompkins, *How to Wreck a Nice Beach*, 181.

50. Rick Davis quoted in the liner notes to Cybotron, *Interface*. See also Reynolds, *Energy Flash*, 8.

51. Tompkins, *How to Wreck a Nice Beach*, 177. See also the following page and Reynolds, *Energy Flash*, 8.

52. In Reynolds, *Energy Flash*, 9–10.

53. Reynolds, *Energy Flash*, 9.

54. Benjamin Noys, *Malign Velocities: Acceleration and Capitalism* (Winchester, UK: Zero Books, 2014), 53. See also Noys, "Dance and Die," 297.

55. Noys, *Malign Velocities*, 94.

56. Noys, "Dance and Die," 298; Noys, *Malign Velocities*, 54. See also note 3 earlier.

57. Noys, *Malign Velocities*, 60.

58. Ibid., 95–96.

59. Ibid., 57.

60. Donna Haraway, *Simians, Cyborgs, and Women: The Reinvention of Nature* (New York: Routledge, 1991), 152. On Haraway and musical cyborgs, including Kraftwerk, see Baumgärtel, *Now and Forever*, 301.

61. Haraway, *Simians, Cyborgs, and Women*, 163.

62. Ibid., 158. For an example of a Marxist-feminist critique from an autonomist perspective see Leopoldina Fortunati, *The Arcane of Reproduction: Housework, Prostitution, Labor and Capital* (1981), trans. Hilary Creek, ed. Jim Fleming (Brooklyn: Autonomedia, 1995).

63. Haraway, *Simians, Cyborgs, and Women*, 166.

64. Ibid., 168.

65. Ibid., 163.

66. Ibid.,153.

67. Ibid., 151.

68. Ibid.

69. Ibid., 152.

70. Ibid., 154. Interestingly, *Enter* and the initial form of Haraway's thinking on this topic (Donna Haraway, "A Manifesto for Cyborgs: Science, Technology, and Socialist Feminism in the 1980s," *Socialist Review* 80 (1985): 65–107) are historical contemporaries.

71. Haraway, *Simians, Cyborgs, and Women*, 150–151.

72. On the right, see, e.g., Francis Fukuyama, *The End of History and the Last Man* (New York: The Free Press, 1992). On the left, see, e.g., Antonio Negri, "Rereading Hegel: The Philosopher of Right," in *Hegel & The Infinite: Religion, Politics, and Dialectic*, trans. Peter Thomas, ed. Slavoj Žižek, Clayton Crockett, and Creston Davis (New York: Columbia University Press), 31–46, especially 40–41.

73. Mirowski, *Machine Dreams*, 441–442.

74. Haraway, *Simians, Cyborgs, and Women*, 152, 153.

75. Fenwick McKelvey and Joshua Neves, "Introduction: Optimization and Its Discontents," *Review of Communication* 21, no. 2 (2021): 106.

76. Haraway, *Simians, Cyborgs, and Women*, 165.

77. Ibid., 158.

78. Marx, *Capital* vol. 1, 129.

79. Seb Franklin, "Cleverness and Drive, or the Cybernetic Fantasy of Value: R.S. Hunt's 'Two Kinds of Work,'" *Grey Room* 68 (Summer 2017): 40. Franklin quotes from R. S. Hunt, "Two Kinds of Work" (~1947; unpublished), Norbert Wiener papers (MC-0022), Box 34B, Folder 1000, Department of Distinctive Collections, MIT Libraries, Cambridge, MA, *iv*.

80. To use the classic cybernetic example of a thermostat, a warm house in a snowstorm, or a cool one in a heat wave, is less probable. Typically, energy will need to have been fed into it to effect these states.

81. Franklin, "Cleverness and Drive," 40.

82. See ibid., 48 and note 18 in the introduction.

83. Franklin, "Cleverness and Drive," 52 (on autonomism, see 49–52).

84. Ibid., 37.

85. Ibid., 45.

86. See note 37 in chapter 2.

87. Marx, *Capital* vol. 1, 509 (see also 312).

88. Ibid., 517. See also note 53 in chapter 2.

89. Rabinbach, *The Human Motor*, 74 (quoting Marx, *Capital* vol. 1, 667).

90. Marx, *Capital* vol. 1, 667.

91. Franklin, "Cleverness and Drive," 54. Marx's discussion similarly relies on gendered labor: "In England women are still occasionally used instead of horses for hauling barges, because the labour required . . . to maintain the women of the surplus population is beneath all calculation" (Marx, *Capital* vol.1, 517).

92. Rabinbach, *The Human Motor*, 59.

93. Ibid., 60.

94. Marx, *Capital* vol. 1, 307.

95. Geoffrey Winthrop-Young and Michael Wutz, "Translators' Introduction" to Kittler, *Gramophone, Film, Typewriter*, xxxiii–xxxiv.

96. Friedrich Kittler, "There Is No Software," *Stanford Literature Review* 9.1 (Spring 1992): 90.

97. Ibid.

98. Haraway, *Simians, Cyborgs, and Women*, 176.

99. Rabinbach, *The Human Motor*, 298.

100. Ibid., 299.

101. As just one example of this fundamental tendency in his thought, see Negri, "Rereading Hegel." Likewise, Cara New Daggett, *The Birth of Energy: Fossil Fuels, Thermodynamics, and the Politics of Work* (Durham, NC: Duke University Press, 2019), which follows Rabinbach in conflating work and labor (83–84), argues that "post-work political movements ease us away from the fever dream of work, highlighting its oppressive and exploitative nature, while potentially inaugurating . . . the kind of radical change called for in the Anthropocene" (206). Many of the environmental critiques of computation examined in chapter 3 share the same orientation.

Conclusion

1. Kittler, "There Is No Software," 82. For his consideration of these definitions of code and their intertwining, see Friedrich Kittler, "Code (or, How You Can Write Something Differently)" (2003), in *Software Studies: A Lexicon*, trans. Tom Morrison and Florian Cramer, ed. Matthew Fuller (Cambridge, MA: MIT Press, 2008), 40–47.

2. Michael Conrad, "The Price of Programmability" in *The Universal Turing Machine: A Half-Century Survey*, ed. Rolf Herken (1988), 2nd ed. (Vienna: Springer-Verlag, 1994/1995), 262. See also note 29 in the introduction and Adam Greenfield, *Radical Technologies: The Design of Everyday Life* (London: Verso, 2017), 21.

3. A well-known example in cryptographic circles, Alexander R. Galloway makes a similar observation (Alexander R. Galloway, *Uncomputable: Play and Politics in the Digital Age* (London: Verso, 2021), 2).

4. See note 96 in chapter 2.

5. Conrad, "The Price of Programmability," 262. In an intellectual sibling to "There Is No Software" distributed via the nettime Internet mailing list, Kittler writes that, in contrast to "hardware . . . the repetition, a million times over, of tiny silicon transistors . . . software is . . . a logical abstraction which, in theory—but only in theory—fundamentally disregards the time and space frameworks of machines in

order to rule them" (Friedrich Kittler, "On the Implementation of Knowledge—Toward a Theory of Hardware," nettime mailing list post, date unknown, archived at http://hydra.humanities.uci.edu/kittler/implement.html, accessed June 14, 2022).

6. See, e.g., the discussion of techno-mediatic metaphors for the mind in Lev Manovich, *The Language of New Media* (Cambridge, MA: MIT Press, 2001), 59–60, Fred Turner, *From Counterculture to Cyberculture: Stewart Brand, the Whole Earth Network, and the Rise of Digital Utopianism* (Chicago and London: University of Chicago Press, 2006), e.g., 15–16, and, more recently and with respect to quantum physics and computation specifically, Jonathan Beller, *The World Computer: Derivative Conditions of Racial Capitalism* (Durham, NC: Duke University Press, 2021), 37.

7. Or, as Kittler's gloss on his disambiguation of encryption from the material world has it, "Condensed into telegraphic style, Turing's statement thus reads: Whether everything in the world can be encoded is written in the stars. But," he continues, "it has by no means been proven that physics, despite its quantum theory, is to be computed solely as a quantity of particles and not as a layering of waves" (Kittler, "Code," 44).

8. John Timmer, "Why Every Quantum Computer Will Need a Powerful Classical Computer," *Ars Technica*, July 9, 2024, https://arstechnica.com/science/2024/07/why-every-quantum-computer-will-need-a-powerful-classical-computer/, accessed August 17, 2024.

9. Lance Fortnow, *The Golden Ticket: P, NP, and the Search for the Impossible* (Princeton: Princeton University Press, 2013), 147. See also note 3 in the introduction. On heuristics see Fortnow, *The Golden Ticket*, 92–97. For a more technical description of quantum computing, see Paul Cockshott, Lewis M. Mackenzie, and Greg Michaelson, *Computation and Its Limits* (Oxford: Oxford University Press, 2012), 128–172.

10. Fortnow, *The Golden Ticket*, 11–28.

11. Wendy Hui Kyong Chun, *Programmed Visions: Software and Memory* (Cambridge, MA: MIT Press, 2010), 9. On Turing's evocation of Laplacean determinism's suitability for computation, again in contradistinction to nature, see 114–115.

12. Ibid., 9.

13. See, e.g., ibid., 47–48. In an earlier essay, she writes, "for computers to become transparency machines, the fact that they compute—that they *generate* text and images rather than merely represent or reproduce what exists elsewhere—must be forgotten" (Wendy Hui Kyong Chun, "On Software, or the Persistence of Visual Knowledge," *Grey Room* 18 (Winter 2004): 27; see also Chun, *Programmed Visions*, 17).

14. Karl Marx and Frederick Engels, *The German Ideology* (1846) Part One, ed. C. J. Arthur (New York: International Publishers, 1970), 47.

15. Roland Barthes, *Mythologies* (1957) trans. Annette Lavers (New York: Noonday Press, 1972), 130.

16. https://x.com/BrunoLatourAIME/status/1245232715835850753. For a broader consideration of the disparagement of abstraction, with specific reference to computer languages, see Andrew Goffey, "Algorithm," in *Software Studies*, ed. Fuller, 19. Despite this, when Goffey emphasizes the "pragmatic functioning" of algorithms, it is less the material circumstances of their execution than, echoing Chun, their "unintended consequences, like the side effects of a program's behavior" (ibid.). Indeed, although algorithms must ultimately be carried out by physical means, he stresses instead the potential for "confusion between the mathematics of algorithms and the physics of real processes, of which Stephen Wolfram's bold speculation that the universe is itself a giant computer is one possible outcome" (16).

17. Anna Kornbluh, *Immediacy, or, The Style of Too Late Capitalism* (London: Verso, 2024), 171. For the full critique see 148–190. In particular, her reading of the methodological rejection of concrete certainty in Hegel and Marx (153–154) correlates with mine in the previous paragraph. *Immediacy* is also notable for acknowledging, paradoxically, the "slow rate of computer-chip and equipment improvements" (27) as one factor among many contributing to the economic torpor lurking behind this high-velocity, anti-mediatory drive. Yet, as Kornbluh ultimately argues, "Hold clogs circulation, keeping some reserve from exchange, or slowing down processing . . . making palpable the work of relation" (196). The onset of 100% utilization may very well, then, signal the end of the immediacy aesthetic.

18. Louis Althusser, "Contradiction and Overdetermination" (1962), in *For Marx* (1965), trans. Ben Brewster (1969) (London: Verso, 2005), 111. See also note 97 in chapter 1. Mark Alizart, *Cryptocommunism*, trans. Robin Mackay (Cambridge, UK: Polity, 2020), links Althusser's intervention to cybernetic models (19–20), further demonstrating how their positions differ from my own.

19. Fredric Jameson, "Cognitive Mapping," in *Marxism and the Interpretation of Culture*, ed. Cary Nelson and Lawrence Grossberg (Urbana, IL: University of Illinois Press, 1988), 354. For a fuller explication of this context see Timothy Scott Brown and Andrew Lison, "Introduction," to *The Global Sixties in Sound and Vision: Media, Counterculture, Revolt*, ed. Timothy Scott Brown and Andrew Lison (New York and Basingstoke, UK: Palgrave Macmillan, 2014), 11n5.

20. Jameson, "Cognitive Mapping," 354. See also Chun, *Programmed Visions*, 73.

21. Louis Althusser, "Ideology and Ideological State Apparatuses (Notes Toward an Investigation)" (1970), in *Lenin and Philosophy and Other Essays*, trans. Ben Brewster (New York: Monthly Review Press, 2001), 118.

22. Richard Dienst, *Still Life in Real Time: Theory After Television* (Durham, NC: Duke University Press, 1994), 141–142.

23. Contrast Marx's famous line, "Men [*sic*] make their own history, but they do not make it just as they please; they do not make it under circumstances chosen by themselves, but under circumstances directly encountered, given and transmitted

from the past" (Karl Marx, *The 18th Brumaire of Louis Bonaparte* (1852/1869/1885) (New York: International Publishers, 1963), 15.

24. See, e.g., Fredric Jameson, "Three Names of the Dialectic" in *Valences of the Dialectic* (London: Verso, 2009), 15.

25. Peter Frase, *Four Futures: Visions of the World after Capitalism* (London: Verso, 2016); Greenfield, *Radical Technologies*, 287–298. As analyses grounded in the Marxist tradition, there will be similarities between our predictions. However, while skeptical, Frase proceeds from "the basic assumption . . . that the trend toward automation will continue in all domains of the economy" (Frase, *Four Futures*, 28), which is precisely what the concept of 100% utilization does not concede. Greenfield's multipronged projections are both similar to Frase's and similarly technology forward.

26. See Jameson, "Three Names of the Dialectic," 46 and note 136 in chapter 5.

27. Theodor W. Adorno, "Why Still Philosophy" (1962), in Theodor W. Adorno, *Critical Models: Interventions and Catchwords*, trans. Henry W. Pickford (1998) (New York: Columbia University Press, 2005), 15.

28. Ibid.

29. As we have already seen in Mark Pesce, "Zombie Moore's Law Shows Hardware Is Eating Software," *The Register*, September 22, 2016, https://www.theregister.com/2016/09/22/the_evolution_of_moores_law_suggests_hardware_is_eating_software/, accessed March 27, 2025, e.g., note 52 in chapter 1.

30. Sean Cubitt, *Finite Media: Environmental Implications of Digital Technologies* (Durham, NC: Duke University Press, 2017), 6. I am also grateful to Jonathan Beller for a discussion which arrived at a similar conclusion.

31. Intel Newsroom, "Intel and Trump Administration Reach Historic Agreement to Accelerate American Technology and Manufacturing Leadership," August 22, 2025, https://newsroom.intel.com/corporate/intel-and-trump-administration-reach-historic-agreement, accessed August 25, 2025. See also note 28 in chapter 2.

32. Min-hee Jung, "Samsung Electronics to Make Its Own CPUs for Smartphones," *Business Korea*, March 6, 2023, https://www.businesskorea.co.kr/news/articleView.html?idxno=110460, accessed January 26, 2025.

33. Omar Sohali, "TSMC Is Rumored to Have Rejected Samsung's Alleged Proposal of Mass Producing the Latter's Exynos Chipsets Using Its Cutting-Edge Lithography," *Wccftech*, January 15, 2025, https://wccftech.com/tsmc-will-not-mass-produce-any-exynos-chipsets-for-samsung/ and "Qualcomm's Snapdragon 8 Elite Gen 2 Will Exclusively Be Mass Produced on TSMC's 3nm 'N3P' Process, with a Rumor Claiming That Samsung Has Lost the Production Bid," *Wccftech*, December 27, 2024, https://wccftech.com/samsung-loses-bid-to-mass-produce-snapdragon-8-elite-gen-2-all-orders-going-to-tsmc/ (both URLs accessed January 25, 2025).

34. Ibid.

35. Andrew Cunningham, "Nvidia GeForce RTX 5090 Costs as Much as a Whole Gaming PC—but It Sure Is Fast," *Ars Technica*, January 24, 2025, https://arstechnica.com/gadgets/2025/01/review-nvidias-geforce-rtx-5090-is-the-first-gpu-that-can-beat-the-rtx-4090/. See also Jarred Walton, "Nvidia's GeForce RTX 5070 at $549—How Does It Stack Up to the Previous Generation RTX 4070?," *Tom's Hardware*, January 7, 2025, https://www.tomshardware.com/pc-components/gpus/nvidias-geforce-rtx-5070-at-usd549-how-does-it-stack-up-to-the-previous-generation-rtx-4070. TSMC's N4 is considered part of its 5nm process (see Taiwan Semiconductor Manufacturing Company, "Logic Technology," https://www.tsmc.com/english/dedicatedFoundry/technology/logic/l_5nm; all URLs accessed January 30, 2025).

36. Angela Yang and Jasmine Cui, "A New AI Assistant from China Has Silicon Valley Talking," *NBC News*, January 27, 2025, https://www.nbcnews.com/tech/tech-news/china-ai-assistant-deepseek-rcna189385, accessed January 30, 2025.

37. As their names suggest, semiconductors have different properties than superconductors, though there have been attempts to reconcile these two technologies, e.g., Heng Wu et al., "The Field-free Josephson Diode in a van der Waals Heterostructure," *Nature* 604 (April 2022): 653–656. For a lay explanation, see Francisco Pires, "Superconducting Breakthrough May Change the Chip Industry Dramatically," *Tom's Hardware*, April 29, 2022, https://www.tomshardware.com/news/superconducting-breakthrough-may-change-the-chip-industry-dramatically, accessed August 6, 2023. See also Martina Trahms et al., "Diode Effect in Josephson Junctions with a Single Magnetic Atom," *Nature* 615 (March 2023): 628–633, in which the authors "investigate the ultimate limits of miniaturization" (628).

38. "If the researchers can devise a way for production to scale up toward millions of Josephson diodes in a single nanoscale chip, all bets are off for the terahertz race" (Pires, "Superconducting Breakthrough"). The technique described in Wu et al., "Field-free Josephson Diode" is based upon superconducting elements that require substantial cooling to function. A room-temperature superconductor would obviate this need, providing further energy savings. For a more skeptical view see Anne Pasek, "Semi-/Superconductors," in *Predictions*, ed. Mél Hogan, Stefan Laser, and Edward Ongweso Jr. (Manchester, UK: Mattering Press, forthcoming).

Index

100% utilization, 42–43, 88, 145, 222n12, 275n17, 276n25. *See also* Moore's law
  of capital, 155–156
  definition, 2–4
  as paradox, 75
  as possibility, 197–200
  and tech culture, 127
  as threat to capitalism, 21–22, 25, 61, 69, 88, 165
13th Floor, The, 8
2008 financial crisis, 59
3dfx, 113
3D XPoint, 34

Abstraction, 6, 9–10, 28, 110, 121, 194
Accelerationism, 4, 42, 149–151, 175–176
Accumulation, 6, 59, 84, 168–169, 181–182, 188, 264n122
  overaccumulation, 156–157
Accuracy, 31, 77, 79–80
Actor-network theory, 194–195
Agamben, Giorgio, 122–123, 253n114
Agency, 160, 193, 196–197
Algorithms, 11, 37, 70, 96, 102, 115, 191–192, 212n31
  in cryptocurrencies, 82–84, 90–91, 95
Alizart, Mark, 169, 236n63, 239n94, 268n25
Alt, Casey, 37–39, 41, 218n80
Althusser, Louis, 15, 195, 196, 221n97
Amdahl, Gene, 111–112
Amdahl's law, 111–112, 116, 153, 201
Anti-unionization, 132–133
Apple, 32, 35–36, 52, 73, 130, 200
Apple Lisa, 73, 231n17
Apple Silicon, 36
Applications (apps), 36, 74, 81, 130, 158
Application-specific integrated circuits (ASICs), 69, 75, 81–83, 94–96, 113
Arbeitskraft, 166–167, 184, 187, 267n12. *See also* Labor power
Arduino, 77
Arrighi, Giovanni, 6
Artificial intelligence (AI), 25, 54, 100–102, 119–122, 245n22, 248n58, 249n64, 250n87. *See also* Neural networks
  and automation, 102, 123, 124
  and cybernetics, 122
  ethics of, 101, 105, 124
  and programming, 107, 121
  software, 103, 115
ASML, 45, 50, 57
Assembly, 53, 76, 132–135, 177
Atkins, Juan, 173–174, 179
Audio technology, 27, 35
  MP3, 62–63, 85
  vocoder, 170, 172–173
Augmented reality, 73

Austerity, 23, 98, 146, 219n89
Automation, 7, 17, 24–25, 35, 43, 66, 124–125, 137, 276n25
  and AI, 102, 118, 124
  computation as, 181
  and labor, 2–4, 64, 118, 137, 182
  in processor manufacture, 53–54, 56, 135
  and unemployment, 137–138, 182, 197
Avanessian, Armen, 150

Back-propagation, 104, 107, 110
Badiou, Alain, 124, 212n31
Barlow, John Perry, 129
Barthes, Roland, 194, 195
Bastani, Aaron, 150
Baudrillard, Jean, 8, 159
Bay Area, San Francisco, 128, 130–131, 147–148. *See also* Silicon Valley
Beckman, Arnold, 137–138, 144, 160, 258n45, 258n51
Beller, Jonathan, 61
Bell Labs, 136, 137, 138, 172
Benanav, Aaron, 151–154, 156, 158, 160–161
Benjamin, Ruha, 146, 260n72–73
Benjamin, Walter, 125
Bergson, Henri, 38
Big data, 128
Binning (microchips), 48, 49
Biopolitics, 144–145
Bitcoin, 19, 83–86, 91–98
  efficiency of, 83, 94
  energy use, 93, 97–98, 237n78, 241n125
Blockchain, 83–84, 90–91, 235n55
Bondyopadhyay, Probir K., 29, 136
Borges, Jorge Luis, 8, 9, 190
Bostrom, Nick, 14–15, 147
Brain, 5, 17, 122, 124, 140, 194, 243n9, 247n43, 251n91
  as parallel processor, 111, 247n47
  simulating, 100, 101, 108–113, 116–117, 120 (*see also* Neural networks)
Brenner, Robert, 152
Brock, David C., 27, 47, 55, 137–138, 147, 210n17, 225n47
Brynjolfsson, Erik, 59, 63–64
Bush, Vannevar, 64–65

C++ (programming language), 79
C (programming language), 11–14, 37
Capital, 23–24, 51, 154–157, 184, 195, 246n38, 256n28
  100% utilization, 155–156
  circuit of, 6
  fixed, 7, 23, 46, 54, 134, 135, 154, 159, 185, 197, 268n25
  surplus, 60
  valorization of, 17, 59, 184
  variable, 7, 23, 46, 54, 92, 134, 154, 197
Capital (Marx), 6, 155, 166, 184, 210n18
Capitalism, 2, 5, 23, 109, 150, 175
  and computing, 5, 59
  digital, 151
  entropy in, 166, 168
  as homeostatic system, 168–169
  postindustrial, 28, 84, 91
  postthermodynamic, 122
  as totality, 197
  and unemployment, 23–24
Carnot, Sadi, 84, 92, 99, 267n12
Causality, 192, 194–195, 251n100, 261n77
CCRU. *See* Cybernetic Culture Research Unit
Central processing units (CPUs), 39–42, 69, 70, 80–82, 113, 216n56, 216n58, 218n82
  Arm-based, 32, 35, 200
  clock speed, 40, 106–107, 115
  and GPUs, 49, 107, 113
Cerebras, 49–50, 52, 56, 100, 200
Ceruzzi, Paul E., 28–29, 43

Chaos, 31, 167, 169, 171, 186–187. *See also* Entropy
ChatGPT, 118. *See also* Artificial intelligence; Neural networks
Chemistry, 45, 105, 138–139, 225n47
Chun, Wendy Hui Kyong, 10–11, 34–35, 43, 79, 96, 103, 106, 121, 146, 193–194, 234n48, 251n91, 251n97
Ciphers, 170–172
Classical programming, 19, 113, 245n29
Clock speed, 40, 106–107
  and neural networks, 111–112
  stagnation, 62, 77, 107, 113, 130, 149
Cloud (computing), 9, 61, 80–81, 96, 101, 128, 130, 255n10
CNNs. *See* Neural networks
Cockshott, Paul, 64, 212n31
Code, 11, 37–38, 76–79, 121, 190. *See also* Programming languages
  efficiency of, 14, 114
  and hardware, 78–79
  human-written, 121, 123
  refactoring, 41, 70
Collectivity, 19, 98–99, 180, 198
Collins, Randall, 24–25, 26
Collis, Avinash, 63–64
Colossus: The Forbin Project, 103, 106
Commodity hardware, 49, 50, 79–80, 106, 112–113
Commodity manufacturing, 63, 65
Communication, 90, 164–165, 168, 170, 187
Communism, 81, 85, 153, 183
Competition, 60, 83–84, 89, 96, 155–156, 183
Compilation, 29, 39, 90–91, 103, 159
Complex instruction set computer (CISC), 112
Complexity, 37, 63, 107, 110
Compression, 90–92, 110, 172, 228n84
Computation, 64, 78, 235n51
  as algorithmic, 37
  as automation, 181
  discrete, 30, 116–117
  efficacy of, 14
  environmental impact, 4–5
  and finance, 59
  and labor, 163, 187, 249n64
  logic of, 29
  materiality of, 4–5, 70, 75, 78, 159
  and nature, 42
  as physical process, 6, 11, 66–67 (*see also* Hardware)
  and production, 5
  and reality, 1, 189–191
  representations of, 11
  as revolution, 8
  semiotics of, 70
  speed of, 64–65
  and thermodynamics, 185
  time of, 14
  as unstable, 67
  work of, 195–196
Computerization, 5, 25, 46, 62–63, 164, 182
Computers
  budget, 14, 35
  as commodities, 118
  complex instruction set computer, 112
  and factories, 64–65
  reduced instruction set computer, 112
Computing
  and capitalism, 5, 59, 129
  economic impacts, 46, 153
  environmental impacts, 87, 97
  and finance, 59–60
  general-purpose, 54
  and labor, 96
  personal, 39
  practical, 64
  and production, 5–7, 62–63
  quantum, 1, 150, 189–192, 195
Concurrency, 39, 108, 114–115, 219n84
Configuration tables, 65, 70

Connectionism, 108–109, 116, 251n97
Conrad, Michael, 15, 30–31, 42–43, 81, 116, 190, 201, 251n97
Consumption, 48–49, 96, 98–99, 152
Convolutional neural networks (CNNs). *See* Neural networks
Cooper, Zane Talley Griffin, 85–88
Costs, 47–52, 58
  in Moore's law, 47–50, 58
COVID-19, 25, 60, 131, 135, 147, 158, 194
CPU. *See* Central processing units
Cramer, Florian, 36
"Cramming More Components onto Integrated Circuits" (Moore), 26–27
Cryptocurrency, 69, 82–89, 97–98. *See also* Bitcoin
  energy use, 97–98, 100, 104
Cryptography, 65, 89–90, 171, 190–191
Cubitt, Sean, 53, 98, 134, 199, 207n39, 217n64, 241n124
Cuboniks, Laboria, 36
Cybernetic Culture Research Unit (CCRU), 149–150, 175, 176
Cybernetics, 21, 109, 122, 137, 164–169, 181, 186–187, 266n6, 268n6
Cyberpunk, 176
Cyborg, 165, 176–180, 186–187
Cybotron, 173–174

Data, 10, 60–63, 104–105, 251n97
  big, 128
  consolidation, 131
  governance, 196
  mining, 97, 102
  value of, 60–61, 84, 104
Data centers, 7, 69, 83, 86–88, 94, 97, 255n10
Davis, Rik, 173–174
Decapsulation, 79, 233n37
DeepSeek R1, 200
Degrowth, 7–8, 98
Deleuze, Gilles, 38, 268n23
"Dematerialization," 7, 21, 64, 66, 67, 71, 154
Democratization, 81, 129
Densification, 49–50
Design, 50–58, 127
  computers in, 53–54
  costs, 52–53
  labor in, 53–54, 57, 134, 224n34
  vs. manufacturing, 45, 50–52, 57–58, 135
Determinism, 16, 22, 55, 121–122, 189, 191–193, 199, 211n26, 252n101
  technological, 47, 99, 151, 160, 198
Detroit, MI, 161, 164, 174, 178
Devs, 1, 8, 189–193, 195–196
Dialectics, 70, 74, 96, 160, 179–180, 195, 197–198
Dienst, Richard, 196
Differentiation, 48, 82
Digital humanities, 42, 219n89
Digital mediation, 2, 8–10, 20, 54, 60–61
Disinformation, 119
Dot-com era, 101, 129–130
Dreyfus, Hubert L., 108–111, 120, 245n29, 251n97

Economics, 145
  of integration, 56–57
  and Moore's law, 17, 47–48, 50, 55, 189
  and technology, 15–16, 55
Economies of scale, 47, 54, 65
Ecosocialism, 8
Education, 117–118
Efficiency, 27, 70, 84–88
  and heat, 88, 92–94
  and meaning, 85, 90–91
  vs. performance, 32
Eglash, Ron, 129, 239n94
Eitel-McCullough, Inc., 132, 133
ELIZA program, 107
Employment, 135. *See also* Unemployment

full, 23, 25, 43, 67, 221n95
underemployment, 24, 177
Emulation, 18, 76, 77, 79–81, 165, 233n39
Encapsulation (object-oriented programming), 39, 43
Energy, 84–86, 166, 181, 184
efficiency, 69
loss, 164
transfer, 93, 182
Engelbart, Douglas, 27–28, 33, 39–40, 73, 211n24, 218n78
Entrepreneurship, 117–118
Entropy, 84–91, 104, 122, 164–169, 171, 186–187, 191, 200, 238n93, 268n25
in capitalism, 124, 166, 168, 182
definition of, 88–89
Environmentalism, 87, 98, 188
Eugenics, 20, 136–139, 144, 146, 246n38
Everett interpretation (physics), 192, 195
Exchange value, 64, 96, 99
Exploits (hacking), 159, 266n135

Fabless firms, 51–52, 57
Fabrication, 45–49. *See also* Manufacturing
capital required, 52–53
and labor, 53, 55, 132–135
and machinery, 55
vertical, 33–34
Facebook, 36, 59, 84, 97, 129–130, 131, 145
Fairchild Semiconductor, 132–133, 140
Feature extractors, 105, 109–110
Feminism, 177, 180
Field-programmable gate arrays (FPGAs), 69, 75–81, 96, 113
Finance, 5–6, 60–62, 67, 97
and computing, 18, 59–61
energy use, 97–98
as immaterial, 59
and Moore's law, 45
and platforms, 61–62
Financialization, 6, 63, 66, 227n71, 228n75
Fixed capital, 23, 46, 54, 134, 135, 154, 159, 197
Flexibility, 109–110, 117, 133–134, 256n28
Flusser, Vilém, 108, 121, 245n29
Fordism, 175
Fortnow, Lance, 191–192
Foxconn, 134
"Fragment on Machines" (Marx), 4
Franklin, Seb, 181–182, 183, 205n18
Frase, Peter, 197, 209n7, 276n25
Full employment, 23, 25, 43, 67, 221n95
Functions (programming), 82

Galloway, Alexander, 9–10, 11, 78, 207n41
Gaming, 18, 48, 69, 76, 79–81, 113, 233n40
Gelsinger, Patrick, 41, 52
Gender, 42, 132, 135, 177
Giedion, Siegfried, 71
Globalization, 6, 24, 59, 109, 133–134, 136
Gmail, 129–130
Goffey, Andrew, 123, 212n31, 275n16
Google, 32, 84, 97, 129, 130, 131, 242n1, 255n12
GPT-2, 119. *See also* Artificial intelligence; Neural networks
Graphical user interface (GUI), 71, 73–75
Graphics processing units (GPUs), 19, 48, 71, 82–83, 95, 103, 113–115, 200, 245n22
and CPUs, 49, 107, 113
and neural networks, 113, 115, 248n60
Greenfield, Adam, 86, 197
Growth, 4, 7–8, 58, 60, 117, 147–148, 161, 168–169

Hall, Gary, 118, 252n107
Halpern, Orit, 167
Handwriting recognition, 105, 112
Haraway, Donna J., 176–180, 186
Hardware, 11, 30–31
  cloning, 77
  customization, 70, 116
  flaws in, 158–159
  and software, 11, 17, 69–75, 78, 96, 193, 232n22, 236n63, 236n71, 249n64
Hardware description languages (HDLs), 76–79
Harvey, David, 59, 62
Hashing (algorithms), 82–84, 89–99
Heat, 86–87, 168, 214n48, 237n76, 238n92
  death, 84, 185, 190
  and efficiency, 88, 92–94
  and work, 86–89, 93, 182–183
Hegel, G. W. F., 43, 104, 194–195, 243n9
Helmholtz, Hermann von, 166, 184
Hillis, W. Daniel, 124, 213n36, 213n38, 216n56, 243n9, 247n43
Hindess, Barry, 195
Hirst, Paul, 195
History, 124–125, 179, 201, 221n97, 253n114, 275n23
Hodges, Andrew, 171
Holgersen, Ståle, 7–8
Homeostasis, 122, 164, 168–169, 176, 184, 187
Huang, Jensen, 107, 116, 248n60
Huhtamo, Erkki, 71
Humanism, 31, 160, 177, 180, 193
Humans
  labor of, 17, 47, 92, 135–136, 151, 163, 184–185, 201
  and machines, 21, 124, 161, 164–165, 172 (*see also* Cyborg)
Hunt, R. S., 181–82
Hwang, Tim, 115, 124, 249n64
Hybridity, 138–140, 144, 179. *See also* Cyborg

Ideology, 15, 109, 121, 140, 194–196
  software as, 78, 121
IDM. *See* Integrated device manufacturer
Image recognition, 104, 107, 117
Incomputability, 5
Individualism, 98, 109, 179
Inefficiency, 86, 94
Inflation, 60
Inform (programming language), 10
Information, 91, 170
  destruction of, 91
  encoded, 190
Informationalization, 64, 153, 183, 191
Information technology, 4, 59, 62, 214n47
Information theory, 90, 92, 172, 181
Innovation, 24–25, 28, 34–37, 157–158, 216n56, 226n57, 232n28, 264n125
Integrated circuits, 1, 25–27, 30, 178, 264n128. *See also* Processors
  and cost, 47–49, 54, 57, 210n17
  cultural impact, 133
  as "immaterial," 178
Integrated device manufacturer (IDM), 51–52, 57–58
Intel 4004 CPU, 30, 40, 82, 212n31
Intel Corporation, 1, 32–33, 34, 48, 113, 136, 199, 224n30, 242n1
  as IDM, 51–52, 57–58
Intellectual property, 29, 258n51
Internet, 29, 59, 97, 128–130

Jameson, Fredric, 6, 24, 46–47, 58, 70, 124, 160, 195
JavaScript, 9, 118, 129
Jevons paradox, 87–88, 95, 237n71
Jones, Rachel, 55

Kay, Alan, 39–40, 72–73, 82, 218n78, 218n80, 231n18
Kelly, Kevin, 148
Kirkwood, Jeffrey West, 83–86, 89, 93, 241n128
Kirschenbaum, Matthew, 5
Kittler, Friedrich, 10, 11, 37, 43, 78, 90, 123, 170, 185, 190, 204n7 243n9
  "There Is No Software," 29–31, 71, 82, 116
Knight, Will, 102, 242n6, 243n9
Knowledge economy, 20, 127, 139, 143, 146
Kornbluh, Anna, 194, 275n17
Kraftwerk, 169–170, 173, 174
Kurzweil, Ray, 147, 148–149, 262n88

Labor, 5, 55–56, 66, 99, 169, 180
  as abstract, 195
  and automation, 2–4, 64, 118, 137, 182
  under communism, 183
  and computation, 64, 96, 163
  costs, 52–53
  dead, 23, 53, 135, 182
  and eugenics, 138–140
  exploitation of, 23, 84, 98, 154, 161
  and fabrication, 55–56, 132–135
  human, 17, 47, 92, 135–136, 151, 163, 184–185, 201
  intensification of, 60–61, 157, 182
  living, 17, 23, 53, 181–182
  and machinery, 46, 56, 66, 99, 135 (*see also* Automation)
  manual, 67, 143–144
  and manufacturing, 50–51, 53–54
  meaning of, 99
  and Moore's law, 55–56
  as negentropic, 169, 183–184
  organized, 25, 137, 145
  postindustrial, 158, 177, 181
  and race, 53, 133–136, 138, 142–145
  rejection of, 187–188
  reproductive, 177, 183
  simulation of, 64
  as social, 180, 182, 184
  as threat to capitalism, 24–25, 186
  white-collar, 25, 132, 143
  and work, 85–86, 92–93, 99–100, 161, 163, 169, 184, 273n101
Labor power, 57, 96, 166–167, 182–184, 187, 197, 264n125
Labor time, 155–156, 169, 180, 183
Language, 90, 103, 106, 108, 121, 123, 196
Large-scale integration (LSI), 72
Late binding, 39
Latour, Bruno, 194
LeCun, Yann, 112
Lécuyer, Christophe, 132
LeNet-5, 112–113, 115
Lithography, 45, 57, 147
Liu, Catherine, 117–118
Logic gates, 7, 76
Lyotard, Jean-François, 9, 195

Machine learning, 19, 101, 106, 115, 128, 248n58, 249n64. *See also* Neural networks
Machines, 54–57
  and humans, 21, 161, 164–165, 172 (*see also* Cyborg)
  and labor, 46, 56, 66, 99, 135 (*see also* Automation)
  and value, 54, 57
Macintosh, 73
Mackay, Robin, 150
macOS, 41
Macroeconomics, 58–59, 155
Majorana, Ettore, 122
Malabou, Catherine, 109, 117, 119, 120, 122, 124, 246n36, 250n87, 251n99, 252n104
Malone, David, 93, 94
Mandel, Ernest, 23, 221n95
Manovich, Lev, 10, 71–72, 74

Manufacturing, 62. *See also* Fabrication
costs, 51, 55
vs. design, 45, 50–52, 57–58, 135
environmental impacts, 133–134
labor of, 50, 53–54, 56
Market, 109
as information processor, 61
segmentation, 48, 51–52
Marks, Laura U., 87
Marx, Karl, 5, 54, 57, 161, 166–167, 180, 183, 194
"Fragment on Machines," 4, 135, 150
Marxism, 3–7, 43, 71, 144–145, 169, 177, 194–195
and technology, 46–47, 268n25
Materialism, 6, 67, 92, 187, 198, 212n31
Matrix, The, 8–9
Matrix multiplication, 113–114
Matroska Multimedia Container, 10
McKelvey, Fenwick, 179, 268n23
Mead, Carver A., 86
Meaning, 89–90, 92, 97
Mechanical Turk, 124–125
Meehan, Mary Beth, 145
Memory, 26, 34, 82, 106, 216n56, 216n58, 251n91
RAM, 82, 111, 159
short-term, 34–35
solid-state, 34, 111, 199
Meritocracy, 143, 146, 161
Micron Technology, Inc., 34
Miniaturization, 27–28, 32, 58, 159, 178
Mirowski, Philip, 61, 165, 168, 179
Misinformation, 16
Modernity, 71, 170
Monopolization, 124, 128–129, 221n98, 227n70
Moore, Gordon E., 1, 26–27, 47, 55–56, 82, 127, 132–133, 136
Moore's law, 113, 220n91
costs in, 47–50, 58
cultural impact, 28–29, 189
definition of, 1–2, 29
as diachronic, 140
and economics, 17, 47–48, 50, 55, 189
end of, 2, 18–21, 33, 188, 196–198, 222n12
introduction of, 26–27, 55, 133
as teleology, 28, 189, 197
and unemployment, 18, 127–128, 197
MP3, 62–63, 85
Multiprocessing. *See* Parallel processing

Nakamura, Lisa, 133–134, 136, 177
Nanotechnology, 147, 227n38
Near (developer), 78–80
Negative value, 63
Negentropy, 168–169, 184
Neoliberalism, 59, 109–110, 124
Neoliberalization, 59, 118
Networks, 16, 61, 87, 101, 104, 216n58
Neural networks, 100, 102–104, 116–117, 125, 161, 251n97, 253n109. *See also* Artificial intelligence
as adaptive, 110
and concurrency, 115
and entropy, 104
and matrix multiplication, 113–114
and Moore's law, 123
and parallel processing, 107–108, 110, 115
as symbol processors, 120–121
training, 100, 104, 120
unknowability of, 102, 104–105, 200, 243n9
Neuroscience, 109–110, 251n99, 252n104
Neves, Joshua, 179
Nik-Khah, Edward, 61
Noble, David F., 117–118
Nonce (cryptocurrency), 83–84, 89, 91
Noyce, Robert, 136
Noys, Benjamin, 164, 175–76, 175–176
NVIDIA, 52–53, 107, 113, 200, 224n32, 242n1, 248n60

Object-oriented programming (OOP), 37–41, 72, 75
Obsolescence, 77, 80, 112–113, 235n51
Offshoring, 24, 53, 127, 132–134
Order, 167–168, 171
Organized labor, 25, 137, 145
Outsourcing, 52, 137, 164

Parallelization, 61, 75, 115–116, 219n84, 248n60
Parallel processing, 39–41, 41, 75, 104–105, 108, 110–112, 130
  and neural networks, 107–108, 110, 115
Parikka, Jussi, 71
Pasek, Anne, 64, 87–88, 93, 255
Performance, 49, 61, 114, 216n58, 219n84, 221n99
  and accuracy, 79–80
  vs. efficiency, 32, 200–201
  and speed, 93, 108, 115
Pesce, Mark, 43, 70, 75, 81
Phenomenology, 28, 108
Physicality, 15, 63, 134
Physics, 93, 163, 173, 190, 197
  quantum, 33–34, 104–105, 121–122, 167, 252n101, 261n77, 274n7
Planck, Max, 88–89
Plasticity, 122
Platforms, 36, 59–62, 84, 118, 128, 130, 157, 196, 197, 227n70
  and finance, 61–62
  and Moore's law, 18, 60
Political economy, 2, 58, 59, 66–67, 123, 174, 185, 196–197
Politics, 124, 169, 196–197
Postcapitalism, 152, 160–161, 183
Posthumanism, 15, 93, 194–195
Postindustrialism, 5–6, 21, 24–25, 164
Postmodernism, 31–32, 138
Poststructuralism, 37–38, 43, 89–90, 90, 187
Prices, 48–49, 72–73, 156
Processing (programming language), 10
Processors
  as absent center, 5
  pricing, 48–49
  product differentiation, 48
  structure, 56–57
  wafer-scale, 49, 50
Production, 54, 57–58, 66, 86, 134, 154
  and computerization, 5–7, 59, 62–63
  forces of, 7, 24, 160, 166
  overproduction, 151–152
  relations of, 64, 67, 150, 160
Productivity paradox, 59, 62–64, 65, 152, 175
Profits, 84, 151, 154–157
  declining, 151, 154, 182, 264n122
  distribution, 58
Programmability, 18, 30–31, 42–43, 81, 96, 119, 146, 193, 212n35
Programming, 65–66, 100, 103
  and AI, 105, 107
  classical, 19, 113
  object-oriented, 37–40, 37–41, 72, 75
Programming languages, 10, 75, 78, 106, 121
  C, 11–14, 37
  C++, 79
  declarative, 37
  Inform, 10
  JavaScript, 9, 118, 129
  Processing, 10
  Python, 11–14, 42, 220n90
  R, 42
Proof-of-stake cryptocurrencies, 95–96
Proof-of-work cryptocurrencies, 83–85, 87, 90, 95–96, 120, 239n94
  as power hungry, 93, 235n51
Python, 11–14, 42, 220n90

Qualcomm, 32, 200
Quantum computing, 1, 150, 189–192, 195
Quantum physics, 33–34, 104–105, 121–122, 167, 252n101, 261n77, 274n7

R (programming language), 42
Race, 53, 133–136, 138, 142–145, 177
Racism, 133, 136–139, 144
RAM. *See* Memory
Raspberry Pi, 14, 35
Rationality, 103–104, 106, 117, 178, 199
Reality, 1, 8, 15, 108, 121–123, 189–191
Recursion, 8–13, 119, 190
Reduced instruction set computer (RISC), 112
Reinforced learning, 111
Remote work, 132
Retrocomputing, 77, 79, 81
Retrogaming, 18, 69, 76
Reynolds, Simon, 174
Robots, 7, 137, 163, 173, 186, 188. *See also* Machines
Rosen, Philip, 15–16
Rowhammer attacks, 159

Samsung, 32, 34, 199–200
Savage, Jon, 173
Self-driving vehicles, 24, 105, 209n4, 214n46, 244n15
Sequential processing, 37–39, 77, 107
Seth, Vikram, 128
Shannon, Claude, 90, 121, 170–171
Sherburne, Philip, 63
Shockley, William, 136–143
Signification, 90, 106
SIGSALY, 170–171
Silicon, 33, 40, 47–49, 69, 106, 134, 189, 225n47, 248n56
  commodity, 82
  custom, 96, 115, 165, 232n28
  special-purpose, 53, 70, 103, 108
Silicon Valley, 25, 36, 127–132, 137, 147
Silicon Valley (series), 110, 260n72
Simulation, 1, 8, 14–15, 46, 64–66, 189–190, 201
  of the brain, 100, 101, 108–113, 116–117, 120 (*see also* Neural networks)
Single-tasking systems, 76
Singularity, 21, 146–148
Skylake processors, 35
Sloterdijk, Peter, 121
Smalltalk, 39–40
Smartphones, 32, 74, 128, 130–131
Socialism, 81, 110, 132, 149, 187
Social media, 36, 129, 131
Software, 18–20, 29–31, 69–81, 91, 101
  and 100% utilization, 200
  and hardware, 11, 17, 69–75, 78, 96, 193, 232n22, 236n63, 236n71, 249n64
  as ideology, 78, 121
  and neural networks, 103, 106, 114–116, 243n9
  "There Is No Software" (Kittler), 29–31, 71, 82, 116
Software Takes Command (Manovich), 71–72, 75
Soldering, 135. *See also* Fabrication; Manufacturing
Speech recognition, 107
Spencer, Matt, 159
Srnicek, Nick, 59–60, 66, 150
Stagnation, 5, 70, 127–128, 151–152, 158, 168–169, 188
  of computing power, 2, 41, 54, 107 (*see also* Moore's law)
  economic, 154, 175–176
Steinberg, Marc, 61
Sterne, Jonathan, 10, 62, 85
Storage, 42, 85. *See also* Memory
  long-term, 34–35
Subjectivity, 15–16
Superintelligence, 20, 148, 152

Surplus value, 117
Surveillance, 97, 105
Sutter, Herb, 41, 43, 114, 115
Symbol manipulation, 69, 100, 108, 120–121
Systems, cybernetic, 164–168, 186
SystemVerilog, 76, 77

Taiwan Semiconductor Manufacturing Company (TSMC), 32, 50, 51, 52, 57, 200, 224n32
Task scheduling, 39, 82
Taylorism, 65, 132
Technological determinism, 47, 99, 151, 189, 191–193
Technology, 92, 150, 160, 173–174, 185, 199
  culture, 127–129
  and economics, 17, 20, 55
  evangelism, 15–16
  and Marxism, 46–47, 268n25
Techno music, 161, 163–164, 173–175
Teleology, 28, 179–180, 197
Terranova, Tiziana, 269n33
Thackray, Arnold, 55
"There Is No Software" (Kittler), 29–31, 71, 82, 116
Thermodynamics, 17, 86, 88–89, 93, 165–166, 181–182, 187–188, 200–201, 239n93, 266n6, 266n7
Thompson, Ken, 159
Time-sharing, 9, 207n32
Toffler, Alvin, 174
Tompkins, Dave, 173
Totalization, 46, 62, 193
TSMC. *See* Taiwan Semiconductor Manufacturing Company
Turing, Alan, 7, 29, 37, 64, 65, 170, 201
Turing machines, 15, 29–31, 70–71, 74, 78, 170, 190–191, 212n31. *See also* Universal machines
Turing test, 108
Turner, Fred, 145, 165

Uber, 24, 36, 59, 118, 158, 209n4
Uncertainty, 9, 33, 110, 167, 200, 242n6
Underemployment, 24, 177
Unemployment, 4, 23–25, 144–145, 150, 174, 261n77
  and automation, 137–138, 182, 197
  and Moore's law, 18, 127–128, 197
Unionization, 25, 132, 142–143
Universal machines, 64, 67, 70, 78. *See also* Turing machines
Unknowability, 104–105, 121–122
User interfaces, 69, 73–74
Use value, 96, 99, 161, 167

Value, 17, 83, 85, 90, 181–184
  creation of, 21, 91–92, 142–143, 201
  of data, 60–61, 84, 104
  distribution of, 97
  exchange value, 64, 96, 99
  and human labor, 47, 184, 201
  judgment, 90–92
  and machinery, 54, 57, 84, 166–167, 181, 184
  negative, 63
  transfer of, 54, 57
  use value, 96, 99, 161, 167
Variable capital, 46, 54, 92, 134, 154, 197
Verilog, 76, 79
Vernam cipher, 170–172
Verstand (Hegel), 28
Villalobos, Ricardo, 163, 174
Virtualization, 9, 92
Virtual reality, 14–15, 73, 147
Vocoder, 170, 172–174
Von Neumann architecture, 106, 159, 216n56

Wages, 23–25, 54, 57–58, 152, 156–157, 236n61
Waldrop, M. Mitchell, 33–34, 43, 75
Waste, 4–5, 83–85, 91–92, 97, 239n100
Web 2.0, 128–130, 255n10

Weil, Simone, 122, 252n101, 253n114
Weizenbaum, Joseph, 107
Welfare state, 23, 24, 124, 144–145
Wiener, Anna, 36
Wiener, Norbert, 64–65, 164–165, 167–168, 172, 181, 266n7, 267n11, 268n23
Williams, Alex, 150, 154, 175
Wilse, Craig, 144–145
Windows (operating system), 41
Wirth's law, 87, 236n71
Work, 86–93, 124, 161, 163–165, 184, 195–196, 253n107
  and automation, 24–25, 175, 186–187
  of computation, 89, 92, 188, 195–196
  and heat, 86–89, 93, 182–183
  and labor, 85–86, 92–93, 99–100, 161, 163, 169, 184, 273n101
  and Moore's law, 47
  rejection of, 187–188
  useful, 89, 92–93
World on a Wire, 8
WSC-3 (processor), 49, 52

Xenofeminism, 36, 42, 134, 135
Xerox Star, 73, 231n18
XML, 129

Yudkowsky, Eliezer, 147

Zeno's paradox, 43, 211n26
Žižek, Slavoj, 121
Zuckerberg, Mark, 145, 261n79
Zuse, Konrad, 65

Publisher contact:
The MIT Press
Massachusetts Institute of Technology
77 Massachusetts Avenue, Cambridge, MA 02139
mitpress.mit.edu

EU Authorised Representative:
Easy Access System Europe, Mustamäe tee 50,
10621 Tallinn, Estonia
gpsr.requests@easproject.com

Printed by Integrated Books International,
United States of America